Empirical Estimates in Stochastic Optimization and Identification

Applied Optimization

Volume 71

Series Editors:

Panos M. Pardalos
University of Florida, U.S.A.

Donald Hearn
University of Florida, U.S.A.

Empirical Estimates in Stochastic Optimization and Identification

by

Pavel S. Knopov
Glushkov Institute of Cybernetics,
Kiev, Ukraine

and

Evgeniya J. Kasitskaya
Glushkov Institute of Cybernetics,
Kiev, Ukraine

KLUWER ACADEMIC PUBLISHERS
DORDRECHT / BOSTON / LONDON

A C.I.P. Catalogue record for this book is available from the Library of Congress.

ISBN 978-1-4419-5224-0

Published by Kluwer Academic Publishers,
P.O. Box 17, 3300 AA Dordrecht, The Netherlands.

Sold and distributed in North, Central and South America
by Kluwer Academic Publishers,
101 Philip Drive, Norwell, MA 02061, U.S.A.

In all other countries, sold and distributed
by Kluwer Academic Publishers,
P.O. Box 322, 3300 AH Dordrecht, The Netherlands.

Printed on acid-free paper

CONTENTS

PREFACE   vii

1 INTRODUCTION   1

2 PARAMETRIC EMPIRICAL METHODS   11
   2.1 Auxiliary Results   12
   2.2 Models with Independent Observations   19
   2.3 Models with Continuous Time   36
   2.4 Models with Restrictions in the Form of Inequalities   44
   2.5 Nonstationary Empirical Estimates   61

3 PARAMETRIC REGRESSION MODELS   71
   3.1 Estimates of the Parameters for Gaussian Regression Models
        with Discrete Time   72
   3.2 Estimates of the Parameters for Gaussian Random Field
        with a Continuous Argument   91
   3.3 Nonstationary Regression Model for Gaussian Field   115
   3.4 Identification of the Parameters for the Stationary Nonlin-
        ear Regression as a Special Case of Stochastic Programming
        Problem   133
   3.5 Nonstationary Regression Model for a Random Field Ob-
        served in a Circle   143
   3.6 Gaussian Regression Models for Quasistationary Random
        Processes   154

4 PERIODOGRAM ESTIMATES FOR
   RANDOM PROCESSES AND FIELDS   163
   4.1 Preliminary Results   163

4.2   Asymptotic Behavior of Periodogram Estimates of the First Type   174

4.3   Asymptotic Behavior of Periodogram Estimates of the Second Type   186

4.4   Periodogram Estimates in $\mathcal{R}^m$ 195

5   **NONPARAMETRIC IDENTIFICATION PROBLEMS**   199

5.1   The Investigation of the General Problem   199

5.2   The Nonparametric Regression Model with Observations in a Finite Number of Curves on the Plane   210

5.3   The Nonparametric Regression Model with Observations in Nodes of a Rectangle   225

5.4   The Periodical Signal Estimation by Observation of Its Mixture with Homogeneous Random Field   230

REFERENCES   239

PREFACE

One of the basic problems of statistical investigation is taking the best in some sense decision by observations of some totality of data. In this book empirical methods for solving of stochastic optimization problems and identification methods closely connected with them are investigated. The main attention is paid to studying of asymptotic behavior of the estimates, proving of the assertions about tending of the considered estimates to optimal ones under unlimited increase of the sample size.

The sufficiently complete idea on empirical methods in the theory of optimization and estimation can be found in the monographs of Ibragimov and Has'minskii [52], Ermoliev and Wets [105], van de Geer [30], Pfanzagl and Wefelmeyer [107] and many others where the new approach to the investigation for discrete and continuous observations is considered.

In the present work some new parametric problems of stochastic optimization and estimation are investigated, the sufficient attention is paid to nonparametric problems and continuous models with a multidimensional argument.

The first chapter is auxiliary. The second one is devoted to investigation of empirical estimates in stochastic optimization problems. In the third chapter parametric regression models are considered, the connection between some of these problems and stochastic optimization problems being studied in the previous chapter is indicated.

The fourth chapter is devoted to studying of so-called periodogram estimates. This class of nonlinear regression models has a number of important practice applications and attracts the attention of specialists for long. In the fifth chapter, which is final, nonparametric problems of stochastic optimization and identification are considered from the general positions, some of these models were not investigated before.

As a rule, the main results of the book consist of obtaining of conditions for strong consistency of the estimates and finding of their asymptotic distribution.

The bibliography does not pretend on completeness, in general the sources being relative directly to the stated material are mentioned.

The authors with feeling of the pleasant duty thank their teachers A.Skorokhod, Yu.Ermoliev, A.Dorogovtsev, M.Yadrenko, their colleagues and collaborators A.Ivanov, Yu.Kozachenko, N.Leonenko, Z.Nekrilova, V.Norkin and many others for fruitful discussion of the results contained in this book.

We are very grateful to Scientific Editor of this book prof. Panos Pardalos and Senior Publishing Editor of the book John Martindale for the valuable advice and collaboration in preparing the manuscript.

1

INTRODUCTION

The following designations will be used: $\mathcal{R}, Q, Z, \mathcal{N}$ are sets of real, rational, integer and natural values respectively; $\mathcal{R}_+$, Z_+ are sets of nonnegative real and integer values; $\mathcal{R}_{++}$ is the set of positive real values. For any abstract sets A_i, $i = \overline{1,m}$ designate $A_1 \times A_2 \times \ldots \times A_m$ the set of all structures $(a_1, a_2, \ldots, a_m)$, where $a_i \in A_i$, $i = \overline{1,m}$; $m \geq 1$. For an arbitrary set A denote $A^m = \underbrace{A \times \cdots \times A}_{m}$. By this way sets $\mathcal{R}^m$, Z^m, $\mathcal{N}^m$, $\mathcal{R}_+^m$, Z_+^m, $\mathcal{R}_{++}^m$ are defined. We denote $\vec{a} = (a_i)_{i=1}^m$ a vector column, and $\vec{a}\,'$ is the corresponding vector-row, $m \geq 1$.

Let

$$\vec{a} = (a_i)_{i=1}^m, \quad \vec{b} = (b_i)_{i=1}^m \in \mathcal{R}^m, \quad m \geq 1.$$

Denote

$$\Pi[\vec{a}, \vec{b}\,[= \left\{ \vec{x} = (x_i)_{i=1}^m \in \mathcal{R}^m \ : \ a_i \leq x_i < b_i, \quad i = 1, m \right\}.$$

Sets $\Pi[\vec{a}, \vec{b}\,]$ and $\Pi]\vec{a}, \vec{b}\,[$ are defined analogously.

For vectors $\vec{T} = (T_i)_{i=1}^m \in \mathcal{R}^m$, $\vec{n} = (n_i)_{i=1}^m \in \mathcal{N}^m$, $m \geq 1$ relations $T_i \to +\infty$, $i = \overline{1,m}$; $n_i \to \infty$, $i = \overline{1,m}$ will be written in such a way:

$$\vec{T} \to \infty, \quad \vec{n} \to \infty.$$

For any $\vec{b} = (b_i)_{i=1}^m \in \mathcal{R}^m$, $A = (a_{jk})_{j,k=1}^m \in \mathcal{M}_m$, $m \geq 1$, where $\mathcal{M}_m$ is the set of matrices $m \times m$ with real elements, denote

$$\| \vec{b} \| = \left(\sum_{i=1}^m (b_i)^2 \right)^{1/2},$$

$$\|A\| = \left(\sum_{j,k=1}^m (a_{jk})^2 \right)^{1/2}. \tag{1.1}$$

Let us consider $\mathcal{R}^m$ and $\mathcal{M}^m$ as vector normalized spaces with norms (1.1).

Now we will introduce a notion of a space with a measure. Suppose that X is some set. It is called a space. For each set $A \subset X$ the set $\overline{A} = X \setminus A$ will be called a complementation of A. An entity of subsets $\mathcal{G}$ of the space X is called σ-algebra if it contains X and has the following properties:

1) if $A \in \mathcal{G}$ then $\overline{A} \in \mathcal{G}$;

2) $\bigcup_i A_i \in \mathcal{G}$ for any countable number of sets $\{A_i\}$ from $\mathcal{G}$.

In this case a pair $(X, \mathcal{G})$ is called a measurable space.

A nonnegative function $\mu = \mu(A)$ on σ-algebra $\mathcal{G}$ (it may be $+\infty$) is called a measure if for any countable number of sets $A_1, A_2, \ldots \in \mathcal{G}$, where $A_i \cap A_j = \emptyset$, $i \neq j$ and $A = \bigcup_i A_i$, we have $\mu(A) = \sum_i \mu(A_i)$. Then the entity $(X, \mathcal{G}, \mu)$ is called a space with a measure. If $\mu(X) < \infty$ the measure μ is said to be finite. If $X = \bigcup_{i \in I} A_i$, where I is a countable set, $\mu(A_i) < \infty$, $i \in I$, then the measure μ is called σ-finite. The measure μ is said to be complete if for any set $A' \subset A \in \mathcal{G}$, $\mu(A) = 0$ we have $A' \in \mathcal{G}$. Then the space $(X, \mathcal{G}, \mu)$ is called complete.

A space with a measure $(\Omega, \mathcal{G}, P)$ is called a probabilistic space if $P(\Omega) = 1$. The measure P is called the probability.

Let $(X, \mathcal{U})$ be some measurable space. The $(\mathcal{G}, \mathcal{U})$ – measurable function $\xi = \xi(w)$, defined on a probabilistic space $(\Omega, \mathcal{G}, P)$, with values in $(X, \mathcal{U})$ is said to be a random variable (a random element) in the fase space $(X, \mathcal{U})$.

Assume that $T \subset \mathcal{R}^m$, $m \geq 1$; $\xi(\vec{t}) = \xi(\vec{t}, \omega)$ is a function of two arguments:

1) a parameter $\vec{t} \in T$;

2) $\omega \in \Omega$, where $(\Omega, \mathcal{G}, P)$ is a probabilistic space; for each fixed $\vec{t} \in T$ $\xi(\vec{t}, \omega)$ is a random variable, defined on the probabilistic space $(\Omega, \mathcal{G}, P)$, in a fase space $(X, \mathcal{U})$.

Then $\xi(\vec{t})$ is called a random field in the fase space $(X, \mathcal{U})$. In the case $m = 1$ a random field is called a random process.

If a random variable $\xi = \xi(\omega)$, defined on a probabilistic space $(\Omega, \mathcal{G}, P)$, with values in $(\mathcal{R}, \mathcal{B}(\mathcal{R}))$ is integrable in Lebesgue sense then the integral

$$E\{\xi\} = \int_{\Omega} \xi(\omega) P(d\omega)$$

is said to be an expectation of ξ.

Let $\vec{\xi} = (\xi_1, \ldots, \xi_n)' \in \mathcal{R}^n$ be a vector random variable on a probabilistic space $(\Omega, \mathcal{G}, P)$. The function $F_{\vec{\xi}}(\vec{x}) = P\{\xi_1 < x_1, \ldots, \xi_n < x_n\}$ of $\vec{x} = (x_1, \ldots, x_n)' \in \mathcal{R}^n$ is called the distribution function of $\vec{\xi}$. The function

$$\varphi_{\vec{\xi}}(\vec{u}) = E\{e^{i(\vec{u}, \vec{\xi})}\}, \quad \vec{u} = (u_1, \ldots, u_n)' \in \mathcal{R}^n$$

is called the characteristic function of $\vec{\xi}$.

A random variable $\vec{\xi} = (\xi_1, \ldots, \xi_n)' \in \mathcal{R}^n$ is said to be Gaussian if its characteristic function is

$$\varphi(\vec{u}) = \exp\left\{ i(\vec{a}, \vec{u}) - \frac{1}{2} \vec{u}' B \vec{u} \right\},$$

where $\vec{a} = (a_k)_{k=1}^n$, $B = (b_{kj})_{k,j=1}^n$ are some vector and matrix respectively. In this case

$$a_k = E\{\xi_k\}, \quad b_{kj} = E\{(\xi_k - a_k)(\xi_j - a_j)\}, \quad k, j = \overline{i, n}.$$

The vector $\vec{a}$ is called the mean of $\vec{\xi}$, and the matrix B – the correlation matrix of $\vec{\xi}$. Gaussian distribution is denoted $\mathcal{N}(\vec{a}, B)$.

A random vector field $\vec{\xi}\,(\vec{t}\,) \in \mathcal{R}^n$, $\vec{t} \in \mathcal{R}^m$, $n, m \geq 1$ is said to be Gaussian if for any $k \in \mathcal{N}$, $\vec{t_1}, \ldots, \vec{t_k} \in \mathcal{R}^m$ the random variable

$$\left(\xi_1(\vec{t}_1), \ldots, \xi_n(\vec{t}_1), \ldots, \xi_1(\vec{t}_k), \ldots, \xi_n(\vec{t}_k) \right)'$$

is Gaussian, where $\vec{\xi}\,(\vec{t}\,) = \left(\xi_1(\vec{t}\,), \ldots, \xi_n(\vec{t}\,) \right)'$.

A random process $\xi(t)$ with values in $(X, \mathcal{U})$, $t \in T \subset \mathcal{R}$ is called stationary in a strict sense if for any u, $t_1, \ldots, t_n$ and $B_1, \ldots, B_n \in \mathcal{U}$

$$P\{\xi(t_1 + u) \in B_1, \ldots, \xi(t_n + u) \in B_n\} = P\{\xi(t_1) \in B_1, \ldots, \xi(t_n) \in B_n\}.$$

Let $\vec{\xi}\,(t) \in \mathcal{R}^p$, $t \in \mathcal{R}$, $p \geq 1$ be an arbitrary random process with

$$E\left\{ \| \vec{\xi}\,(t) \|^2 \right\} < \infty$$

for any t. Then the matrix function

$$R(t, s) = E\left\{ \left(\vec{\xi}\,(t) - E\{\vec{\xi}\,(t)\} \right) \left(\vec{\xi}\,(s) - E\{\vec{\xi}\,(s)\} \right)' \right\}$$

is defined and called the correlation function of $\vec{\xi}\,(t)$. If $E\left\{ \vec{\xi}\,(t) \right\} = \vec{a}$ is a constant and $R(t, s) = R_1(t - s)$ depends only on $t - s$ then the random process $\vec{\xi}\,(t)$ is called stationary in a wide sense.

A random field $\xi(\vec{t}\,)$, $\vec{t} \in \mathcal{R}^m$, $m \geq 1$, defined on a probabilistic space $(\Omega, \mathcal{G}, P)$, with values in $(X, \mathcal{U})$ is called homogeneous in a strict sense if for any $\vec{u} \in \mathcal{R}^m$, $n \in \mathcal{N}$, $\vec{t}_1, \ldots, \vec{t}_n \in \mathcal{R}^m$, $B_1, \ldots, B_n \in \mathcal{U}$

$$P\left\{ \xi(\vec{t}_1 + \vec{u}) \in B_1, \ldots, \xi(\vec{t}_n + \vec{u}) \in B_n \right\} = P\left\{ \xi(\vec{t}_1) \in B_1, \ldots, \xi(\vec{t}_n) \in B_n \right\}$$

A random field $\vec{\xi}\,(\vec{t}\,) \in \mathcal{R}^p$, $\vec{t} \in \mathcal{R}^m$, $p, m \geq 1$ with $E\left\{ \| \vec{\xi}\,(\vec{t}\,) \|^2 \right\} < \infty$ is said to be homogeneous in a wide sense if $E\left\{ \vec{\xi}\,(\vec{t}\,) \right\} = \vec{k} = const$ and for the correlation function we have $R(\vec{t}, \vec{s}) = R_1(\vec{t} - \vec{s})$. In this case

$$R_1(\vec{t} - \vec{s}) = \int_{\mathcal{R}^m} e^{i(\vec{\lambda}, \vec{t} - \vec{s})} F(d\vec{\lambda}),$$

where $F(\mathcal{R}^m) = R_1(\vec{0})$, $F(\cdot)$ is a finite matrix measure on $(\mathcal{R}^m, \mathcal{B}(\mathcal{R}^m))$. If

$$F(\Delta) = \int_{\Delta} f(\vec{\lambda}) d\vec{\lambda}, \quad \Delta \in \mathcal{B}(\mathcal{R}^m)$$

then $f(\vec{\lambda})$ is called the spectral matrix density of $\vec{\xi}(\vec{t})$. A function $F(\vec{\lambda}) = F\left(\Pi\,]-\infty, \vec{\lambda}\,[\right)$ is said to be the spectral function of $\vec{\xi}(\vec{t})$.

A linear transformation g of $\mathcal{R}^m$ which does not change its orientation and the distance from any point to $\vec{0}$ $\left(\|g\,\vec{t}\,\| = \|\,\vec{t}\,\|\right)$ is called a turn. A random field $\vec{\xi}(\vec{t})$, $\vec{t} \in \mathcal{R}^m$ on $(\Omega, \mathcal{G}, P)$ with values in $(X, \mathcal{U})$ is said to be isotropic if for any turn g in $\mathcal{R}^m$ and each $\vec{t}_1, \ldots, \vec{t}_n \in \mathcal{R}^m$, $B_1, \ldots, B_n \in \mathcal{U}$

$$P\left\{\xi(g\,\vec{t}_1) \in B_1, \ldots, \xi(g\,\vec{t}_n) \in B_n\right\} = P\left\{\xi(\vec{t}_1) \in B_1, \ldots, \xi(\vec{t}_n) \in B_n\right\}.$$

If for a stationary in a strict sense random process with a discrete (continuous) parameter $\{\xi_i,\ i \in Z\}$ $(\{\xi(t),\ t \in \mathcal{R}\})$ on $(\Omega, \mathcal{G}, P)$ with values in $(X, \mathcal{U})$ we have

$$\sup_{\substack{A_1 \in \mathcal{F}_{-\infty}^i \\ A_2 \in \mathcal{F}_{i+k}^{+\infty}}} \left| P(A_1 \bigcap A_2) - P(A_1) P(A_2) \right| = \alpha(k) \to 0, \quad k \to \infty$$

$$\left(\sup_{\substack{A_1 \in \mathcal{F}_{-\infty}^t \\ A_2 \in \mathcal{F}_{t+\tau}^{+\infty}}} \left| P(A_1 \bigcap A_2) - P(A_1) P(A_2) \right| = \alpha(\tau) \to 0, \quad \tau \to \infty \right),$$

where $\mathcal{F}_k^j = \sigma\{\xi_i,\ k \le i \le j\}$ $(\mathcal{F}_a^b = \sigma\{\xi(t),\ t \in [a, b]\})$ – the minimal σ-algebra of subsets of Ω, containing all sets $\{\omega\ :\ \xi_i(\omega) \in B\},\ k \le i \le j$

$(\{\omega \ : \ \xi(t,\omega) \in B\}, \ t \in [a,b])$, $B \in \mathcal{U}$, then the process ξ_i $(\xi(t))$ is said to satisfy a strong mixing condition with the coefficient $\alpha(k)$ $(\alpha(\tau))$.

Suppose that for a homogeneous in a strict sense random field $\xi(\vec{t}\,)$, $\vec{t} \in \mathcal{R}^m$, $m \geq 1$, on $(\Omega, \mathcal{G}, P)$ with values in $(X, \mathcal{U})$ there exists such a function $\Psi(d)$, $d \geq 0$, $\Psi(d) \searrow 0$, $d \to \infty$ that for any sets S_1, $S_2 \subset \mathcal{R}^m$

$$\sup_{\substack{A_1 \in \mathcal{F}(S_1) \\ A_2 \in \mathcal{F}(S_2)}} \left| P(A_1 \textstyle\bigcap A_2) - P(A_1)P(A_2) \right| \leq \Psi\left(d(S_1, S_2) \right),$$

where

$$\mathcal{F}(S) = \sigma \Big\{ \xi(\vec{t}\,), \quad \vec{t} \in S \Big\},$$

$$d(S_1, S_2) = \inf \Big\{ \| \vec{t}_1 - \vec{t}_2 \|, \quad \vec{t}_1 \in S_1, \quad \vec{t}_2 \in S_2 \Big\}.$$

Then the field $\xi(\vec{t}\,)$ is said to satisfy a strong mixing condition with the coefficient $\Psi(d)$.

Let ξ_n, $n \geq 1$ and ξ be random variables on $(\Omega, \mathcal{G}, P)$ with values in the metric space $\Big(X, \mathcal{B}(X) \Big)$. The sequence ξ_n is said to be convergent to ξ with probability 1, $n \to \infty$ if

$$P\Big\{ \omega \ : \ \xi_n(\omega) \to \xi(\omega), \quad n \to \infty \Big\} = 1.$$

If for any $\varepsilon > 0$

$$P\Big\{ \omega \ : \ \rho\left(\xi_n(\omega), \xi(\omega) \right) > \varepsilon \Big\} \to 0, \qquad n \to \infty,$$

where $\rho(\cdot)$ – a metric in X, then ξ_n is said to be convergent to ξ in probability.

Assume that $\vec{\xi}_n$, $n \geq 1$ and $\vec{\xi}$ are random variables in $\mathcal{R}^m$, $m \geq 1$. The sequence $\vec{\xi}_n$ is said to be convergent to $\vec{\xi}$ weakly or in distribution if for all $\vec{x} \in \mathcal{R}^m$

$$F_{\vec{\xi}_n}(\vec{x}) \to F_{\vec{\xi}}(\vec{x}), \quad n \to \infty,$$

where $F_{\vec{\xi}}(\vec{x})$ is a distribution function, or if for any $\vec{u} \in \mathcal{R}^m$

$$\varphi_{\vec{\xi}_n}(\vec{u}) \to \varphi_{\vec{\xi}}(\vec{u}), \quad n \to \infty,$$

where $\varphi_{\vec{\xi}}(\vec{u})$ is a characteristic function.

The following ergodic theorems take place.

Theorem 1 *[91] Let* $\left\{\vec{\xi}_i, i \geq 1\right\}$ *be a sequence of independent identically distributed random variables,* $E\{\vec{\xi}_i\} = \vec{a}$. *Then with probability 1*

$$\frac{1}{n} \sum_{i=1}^{n} \vec{\xi}_i \to \vec{a}, \quad n \to \infty.$$

This theorem is called a law of large numbers.

Suppose that $\left\{\xi_i, i \in Z\right\}$ $\left(\left\{\xi(t), t \in \mathcal{R}\right\}\right)$ is a stationary in a strict sense random process with a discrete (continuous) parameter and values in $(X, \mathcal{U})$. It is called ergodic or metrically transitive [91] if for any measurable function $\varphi : X \to \mathcal{R}$ with $E\left\{|\varphi(\xi_i)|\right\} < \infty$ $\left(E\left\{\left|\varphi\big(\xi(t)\big)\right|\right\} < \infty\right)$ we have

$$P\left\{\frac{1}{n}\sum_{i=1}^{n}\varphi(\xi_i) \to E\{\varphi(\xi_1)\}, \quad n \to \infty\right\} = 1,$$

$$\left(P\left\{\frac{1}{T}\int_0^T \varphi(\xi(t))\,dt \to E\{\varphi(\xi(0))\}, \quad T \to \infty\right\} = 1\right).$$

A random field $\xi(\vec{t})$, $\vec{t} \in T \subset \mathcal{R}^m$, $m \geq 1$ on $(\Omega, \mathcal{G}, P)$ with values in $(X, \mathcal{U})$ is said to be measurable if the function $\xi(\vec{t}, \omega) : T \times \Omega \to X$ is measurable. The field is called separable relatively to a set $I \subset T$ if I is countable and dense in T, and there exists such a set $N \in \mathcal{G}$, $P(N) = 1$ that for any ball

$$v(r) = \left\{\vec{t} \in \mathcal{R}^m : \|\vec{t}\| \leq r\right\}$$

we have

$$\left\{\omega : \sup_{\vec{t} \in I \bigcap v(r)} \xi(\vec{t}) = \sup_{\vec{t} \in T \bigcap v(r)} \xi(\vec{t})\right\} \supset N,$$

$$\left\{ \omega : \inf_{\vec{t}\in I\bigcap v(r)} \xi(\vec{t}) = \inf_{\vec{t}\in T\bigcap v(r)} \xi(\vec{t}) \right\} \supset N.$$

A field $\xi(\vec{t}\,) \in \mathcal{R}$, $\vec{t}\in \mathcal{R}^m$ is said to be mean square continuous if for each $\vec{t}$, $\vec{t}_n \to \vec{t}$, $n \to \infty$ we have

$$E\left\{ \left(\xi(\vec{t}_n) - \xi(\vec{t}\,) \right)^2 \right\} \to 0, \quad n \to \infty.$$

Theorem 2 [78] *Let $\xi(\vec{t}\,) \in \mathcal{R}$, $\vec{t}\in \mathcal{R}^m$, $m \geq 1$ be a homogeneous in a strict sense real random field, which is mean square continuous, measurable and separable. Suppose that it satisfies a strong mixing condition with*

$$\Psi(d) \leq \frac{c}{1 + d^{m+\varepsilon}}, \quad d \geq 0; \quad 0 < c < +\infty, \quad \varepsilon > 0.$$

Assume that $E\left\{ \xi(\vec{t}\,) \right\} = 0$ and $E\left\{ \left| \xi(\vec{t}\,) \right|^{4+\delta} \right\} < \infty$ for some $\delta > 4m/\varepsilon$. Then with probability 1

$$\frac{1}{T^m} \int_{[0,T]^m} \xi(\vec{t}\,) d\vec{t} \to 0, \quad T \to \infty.$$

There are central limit theorems.

Theorem 3 [91] *Let $\vec{\xi}_i \in \mathcal{R}^p$, $i \geq 1$ be a sequence of independent identically distributed random variables, $E\{\vec{\xi}_i\} = \vec{a}$, $E\left\{ (\vec{\xi}_i - \vec{a})(\xi_i - \vec{a})' \right\} = R = (r_{kj})_{k,j=1}^p$. Then*

$$\frac{1}{\sqrt{n}} \sum_{i=1}^n (\vec{\xi}_i - \vec{a}) \Longrightarrow \mathcal{N}(\vec{0}, R), \quad n \to \infty,$$

where "$\Longrightarrow$" means the weak convergence, $\mathcal{N}(\cdot, \cdot)$ is Gaussian distribution.

Theorem 4 [91] *Suppose that $\left\{ \vec{\xi}_i, i \in Z \right\} \left(\left\{ \vec{\xi}(t), t \in \mathcal{R} \right\} \right)$, is a stationary in a strict sense random process with a discrete (continuous) parameter and values in $\mathcal{R}^m$, $m \geq 1$, satisfying a strong mixing condition with $\alpha(k) = O(k^{-1-\varepsilon})$,*

$k \to \infty \ (\alpha(\tau) = O(\tau^{-1-\varepsilon}), \ \tau \to \infty), \ \varepsilon > 0.$

Assume that $E\left\{ \vec{\xi}_i \right\} = \vec{0}, \ \left(E\left\{ \vec{\xi}(t) = \vec{0} \right\} \right),$

$E\left\{ \| \vec{\xi}_i \|^{2+\delta} \right\} < \infty \ \left(E\left\{ \| \vec{\xi}(t) \|^{2+\delta} \right\} < \infty \right) \ for \ some \ \delta > 4/\varepsilon,$

and $\det g(0) \neq 0$, *where* $g(\lambda)$ *is a spectral density matrix of the process* $\vec{\xi}_i \ \left(\vec{\xi}(t) \right)$.

Then
$$\frac{1}{\sqrt{n}} \sum_{i=1}^{n} \vec{\xi}_i \Longrightarrow \mathcal{N}\left(\vec{0}, 2\pi \, g(0) \right), \quad n \to \infty.$$

$$\left(\frac{1}{\sqrt{T}} \int_0^T \vec{\xi}(t) \, dt \Longrightarrow \mathcal{N}\left(\vec{0}, 2\pi \, g(0) \right), \quad T \to \infty \right).$$

Theorem 5 *[61] Let* $\vec{\xi}(\vec{t}) \in \mathcal{R}^p$, $t \in \mathcal{R}^m$, $p \geq 1$, $m \geq 1$ *be a homogeneous in a strict sense random field, which traectories are continuous with probability 1. Suppose that* $\vec{\xi}(\vec{t})$ *satisfies the strong mixing condition with*

$$\Psi(d) = O\left(d^{-m-\varepsilon} \right), \quad d \to \infty; \quad \varepsilon > 0$$

and

$$E\left\{ \vec{\xi}(\vec{t}) \right\} = \vec{0}, \quad E\left\{ \| \vec{\xi}(\vec{t}) \|^{2+\delta} \right\} < \infty$$

for some $\delta > 2m/\varepsilon$. *Assume that* $\det g(\vec{0}) \neq 0$, *where* $g(\vec{\lambda})$ *is a matrix of the spectral density of* $\vec{\xi}(\vec{t})$. *Then*

$$\frac{1}{T^{m/2}} \int_{[0,T]^m} \vec{\xi}(\vec{t}) \, d\vec{t} \Longrightarrow \mathcal{N}\left(\vec{0}, (2\pi)^m g(\vec{0}) \right), \quad T \to \infty.$$

2

PARAMETRIC EMPIRICAL METHODS

In this chapter some variants of stochastic programming problems are considered. Three cases are investigated:

1) the random factor in the problem is represented by a random element from some metric space, and empirical estimates of the criterion function are made by independent observations of the random element;

2) the random factor is a stationary in a strict sense ergodic random process with a discrete parameter, and values of the process in the finite number of points are observed;

3) the random factor is a measurable stationary in a strict sense ergodic random process with a continuous parameter, and the part of a realization of the process is observed.

It is supposed that the solution belongs to the closed subset of the Euclidian finite-dimensional space, and that it is unique.

Instead of the original criterion functions the empirical functions are minimized.

It is proved that under rather general conditions minimum points of the empirical functions converges to the solution of the original problem with probability 1.

The grate attention is paid to investigation of the asymptotic distribution of the estimates. It is shown that in general it is not normal, it will be normal only if the minimum point of the former criterion function is internal.

2.1 AUXILIARY RESULTS

Lemma 1 *The following results will be necessary for our research. Let X be an arbitrary measurable in Lebesgue sense subset of $\mathcal{R}^l$, $l \geq 1$; $(Y, \mathcal{Q}, \nu)$ is some space with a finite measure or σ-finite one; $h : X \times Y \to \mathcal{R}$ is a function continuous in the first argument for any fixed $y \in Y \setminus Y'$, $\nu(Y') = 0$, and ν-measurable in the second argument for each $\vec{x} \in X$. Introduce a space with a measure $\left(X, \mathcal{B}(X), \mu \right)$, where μ is a Lebesgue measure on $\mathcal{B}(X)$.*

Then the function h is $\tilde{\sigma}\left\{ \mathcal{B}(X) \times \mathcal{Q} \right\}$ – measurable and if $Y' = \emptyset$ then h is $\sigma\left\{ \mathcal{B}(X) \times \mathcal{Q} \right\}$– measurable.

Proof. Let $\vec{1}$ be a vector from $\mathcal{R}^l$ with all coordinates equal to 1. For any $n \in \mathcal{N}$, $\vec{\jmath} \in Z^l$ denote

$$X(n, \vec{\jmath}\,) = \Pi\,]\,\frac{\vec{\jmath}}{n}, \frac{\vec{\jmath} + \vec{1}}{n}\,[\, \bigcap X$$

and if the set $X(n, \vec{\jmath}\,)$ is not empty then fix a point $\vec{a}\,(n, \vec{\jmath}\,) \in X(n, \vec{\jmath}\,)$ arbitrarily. Then for any pair $n \in \mathcal{N}$, $\vec{x} \in X$ consider the vector $\vec{\jmath} = \vec{\jmath}\,(n, \vec{x}\,) \in Z^l$ for which $\vec{x} \in X(n, \vec{\jmath}\,)$.

Introduce the function

$$h_n(\vec{x}, y) = h\left(\vec{a}\,\left(n, \vec{\jmath}\,(n, \vec{x}\,) \right), y \right);\quad n \in \mathcal{N},\quad \vec{x} \in X,\quad y \in Y.$$

It will be shown that for each $n \in \mathcal{N}$ the mapping $h_n(\vec{x}, y)$ is $\sigma\{\mathcal{B}(X) \times \mathcal{Q}\}$ – measurable. Fix $n \in \mathcal{N}$ and a set $A \in \mathcal{B}(\mathcal{R})$. Then

$$\left\{ (\vec{x}, y) \in X \times Y : h_n(\vec{x}, y) \in A \right\} =$$

$$= \bigcup_{\vec{\jmath} \in Z^l} \left\{ (\vec{x}, y) \in X \times Y : h_n(\vec{x}, y) \in A,\quad \vec{x} \in X(n, \vec{\jmath}\,) \right\} =$$

$$= \bigcup_{\vec{\jmath} \in Z^l} \left\{ (\vec{x}, y) \in X \times Y : h\left(\vec{a}\,(n, \vec{\jmath}\,), y \right) \in A,\quad \vec{x} \in X(n, \vec{\jmath}\,) \right\} =$$

$$= \bigcup_{\vec{\jmath} \in Z^l} \left(X(n, \vec{\jmath}) \times \left\{ y \in Y : h\left(\vec{a}\,(n, \vec{\jmath}), y \right) \in A \right\} \right).$$

For any $\vec{\jmath} \in Z^l$ $X(n, \vec{\jmath}) \in \mathcal{B}(X)$. Clearly,

$$\Pi \left[\frac{\vec{\jmath}}{n}, \frac{\vec{\jmath} + \vec{1}}{n} \right[= \bigcap_{m=1}^{\infty} \Pi \left] \frac{\vec{\jmath}}{n} - \frac{\vec{1}}{m}, \frac{\vec{\jmath} + \vec{1}}{n} \right[,$$

hence

$$X(n, \vec{\jmath}) = \bigcap_{m=1}^{\infty} \left(\Pi \left] \frac{\vec{\jmath}}{n} - \frac{\vec{1}}{m}, \frac{\vec{\jmath} + \vec{1}}{n} \right[\bigcap X \right).$$

Since the function h is measurable in the second argument and the set Z^l is discrete, the function $h_n(\vec{x}, y)$ is $\sigma\{\mathcal{B}(X) \times \mathcal{Q}\}$ – measurable.

Then for all $\vec{x} \in X$, $n \in \mathcal{N}$, $\vec{\jmath} = \vec{\jmath}\,(n, \vec{x})$

$$\vec{a}\,(n, \vec{\jmath}), \ \vec{x} \in \Pi \left[\frac{\vec{\jmath}}{n}, \frac{\vec{\jmath} + \vec{1}}{n} \right[.$$

Hence

$$\left\| \vec{a}\,(n, \vec{\jmath}\,(n, \vec{x})) - \vec{x} \right\| < \left((1/n)^2 l \right)^{1/2} = \sqrt{l}/n.$$

Then

$$\vec{a}\,(n, \vec{\jmath}\,(n, \vec{x})) \to \vec{x}, \quad n \to \infty.$$

This implies for any $\vec{x} \in X$, $y \in Y \setminus Y'$ that

$$h_n(\vec{x}, y) \to h(\vec{x}, y), \quad n \to \infty.$$

Now to prove the lemma it is sufficiently to check if $g(X \times Y') = 0$, where g is the product of measures μ and ν.

Evidently,

$$X = \bigcup_{m=1}^{\infty} \left(\Pi \left] -\vec{m}, \vec{m} \right[\bigcap X \right),$$

where $\vec{m} \in \mathcal{R}^l$, $\vec{m} = (m)_{k=1}^l$, $m \in \mathcal{N}$. Hence

$$X \times Y' = \bigcup_{m=1}^{\infty} \left((\Pi \,] - \vec{m}, \; \vec{m} \; [\; \bigcap X) \times Y' \right).$$

Then

$$\Pi \,] - \vec{m}, \; \vec{m} \; [\; \bigcap X \in \mathcal{B}(X),$$

$$\left(\Pi \,] - \vec{m}, \vec{m} \; [\; \bigcap X \right) \times Y' \subset \left(\Pi \,] - (\overrightarrow{m+1}), \overrightarrow{m+1} \; [\; \bigcap X \right) \times Y', \quad m \in \mathcal{N}.$$

The continuity of the measure implies

$$q(X \times Y') = \lim_{m \to \infty} q \left((\Pi \,] - \vec{m}, \; \vec{m} \; [\; \bigcap X) \times Y' \right) =$$

$$= \lim_{m \to \infty} \left(\mu \left(\Pi \,] - \vec{m}, \vec{m} \; [\; \bigcap X \right) \times \nu (Y') \right) = 0.$$

Lemma 2 [118] Let $\left\{ \vec{\xi}_i, \; i \in \mathcal{N} \right\}$, $\left\{ \vec{\eta}_i, \; i \in \mathcal{N} \right\}$ be sequences of random vector variables; $\vec{\xi}_i \in \mathcal{R}^k$, $\vec{\eta}_i \in \mathcal{R}^l$; $k \geq 1$, $l \geq 1$; $\vec{\xi}_i$ converges in distribution to $\vec{\xi}$, $\vec{\eta}_i$ converges by probability to $\vec{c}$, $i \to \infty$; $\vec{\xi}$ is some random variable, $\vec{c}$ is some vector. If $\vec{\varphi}$ is the continuous mapping from $\mathcal{R}^{k+l}$ to $\mathcal{R}^m$, $m \geq 1$ then $\vec{\varphi}(\vec{\xi}_i, \vec{\eta}_i)$ converges in distribution to $\vec{\varphi}(\vec{\xi}, \vec{c})$, $i \to \infty$.

Lemma 3 Let X be an arbitrary subset of some separable metric space with a metric ρ, $(Y, \mathcal{Q})$ be a measurable space, $f = f(x, y) : X \times Y \to \mathcal{R}$ be a function, continuous in the first argument for any y and $\mathcal{Q}$ – measurable in the second argument for each x. Then the mappings

$$g(y) = \inf_{x \in X} f(x, y), \quad h(y) = \sup_{x \in X} f(x, y), \quad y \in Y,$$

are $\mathcal{Q}$-measurable.

Proof. Let X' be a discrete dense everywhere subset of X. The properties of measurable functions imply that the mapping $g_1(y) = \inf\limits_{x \in X'} f(x, y)$, $y \in Y$ is $\mathcal{Q}$ - measurable. Let us fix an arbitrary element $y \in Y$. It will be shown that $g(y) = g_1(y)$. It is sufficiently to prove that $f(x, y) \geq g_1(y)$, $x \in X$. Fix $x \in X$. There exists a sequence $\{x_n\}$ of elements from X', converging to x, $n \to \infty$. Since f is continuous in the first argument, $f(x_n, y) \to f(x, y)$, $n \to \infty$. Then $f(x_n, y) \geq g_1(y)$, $n \in \mathcal{N}$. Hence $f(x, y) \geq g_1(y)$. Then the function $g(y) = g_1(y)$, $y \in Y$ is $\mathcal{Q}$ - measurable. Then

$$h(y) = - \inf\limits_{x \in X} (-f(x, y)), \quad y \in Y.$$

The same arguments as mentioned above can be applied to the function $-f$. Hence the mapping $h(y)$, $y \in Y$ is $\mathcal{Q}$ - measurable.

The following lemma is a variation of Fatu lemma.

Lemma 4 *Let $(X, \mathcal{X}, \mu)$ be a space with a finite measure, $\mu(X) > 0$; $\{f_n = f_n(x) : X \to \mathcal{R}, n \in \mathcal{N}\}$ is a sequence of nonnegative $\mathcal{X}$ - measurable functions. Suppose that μ-almost everywhere on X $f_n(x) \to \infty$, $n \to \infty$. Then*

$$\int\limits_X f_n(x) \, d\mu \to \infty, \quad n \to \infty.$$

Theorem 6 *[106] Let T be an arbitrary closed or open subset of $\mathcal{R}^l$, $l \geq 1$; $(X, \mathcal{U})$ – some measurable space. Suppose that $f : T \times X \to [-\infty, \infty]$ is a function satisfying the conditions:*

1) $f(t, x)$, $t \in T$ is continuous for all $x \in X$;

2) $f(t, x)$, $x \in X$ is $\mathcal{U}$ - measurable for each $t \in T$;

3) for any $x \in X$ there exists $t^* \in T$ with $f(t^*, x) = \inf\limits_{t \in T} f(t, x)$.

Then there exists the measurable mapping $\varphi : X \to T$ with

$$f(\varphi(x), x) = \inf\limits_{t \in T} f(t, x), \quad x \in X.$$

Theorem 7 [15] *Let $(\Omega, \mathcal{U}, P)$ be a complete probabilistic space and K be a compact subset of some Banach space with a norm $\|\cdot\|$. Suppose that*

$$\left\{ \mathcal{U}_{\vec{T}}, \quad \vec{T} \in \mathcal{R}^{m}_{++}(\mathcal{N}^m) \right\}$$

is the family of σ-algebras such that $\mathcal{U}_{\vec{T}} \subset \mathcal{U}, \mathcal{U}_{\vec{T}} \subset \mathcal{U}_{\vec{S}}, \vec{T} < \vec{S}$ (each component of $\vec{T}$ is less than the corresponding component of $\vec{S}$), and

$$\left\{ Q_{\vec{T}}(s) = Q_{\vec{T}}(s,\omega) : (s,\omega) \in K \times \Omega, \quad \vec{T} \in \mathcal{R}^{m}_{++}(\mathcal{N}^m) \right\}$$

is the family of real functions satisfying the following conditions:

1) for fixed $\vec{T}$ and ω the function $Q_{\vec{T}}(s,\omega)$, $s \in K$, is continuous;

2) for fixed $\vec{T}$ for each $s \in K$ the function $Q_{\vec{T}}(s,\omega)$ is $\mathcal{U}_{\vec{T}}$–measurable;

3) for some element $s_0 \in K$ for each $s \in K$

$$P\left\{ \lim_{\vec{T} \to \infty} Q_{\vec{T}}(s,\omega) = \Phi(s;s_0) \right\} = 1,$$

where $\Phi(s;s_0)$, $s \in K$ is the real function, which is continuous on K and satisfies the condition

$$\Phi(s;s_0) > \Phi(s_0;s_0), \quad s \neq s_0;$$

4) for any $\delta > 0$ there exist $\gamma_0 > 0$ and the function $c(\gamma)$, $\gamma > 0$, $c(\gamma) \to 0$, $\gamma \to 0$ such that for any element $s' \in K$ and any $\gamma : 0 < \gamma < \gamma_0$

$$P\left\{ \varlimsup_{\vec{T} \to \infty} \sup_{\substack{\{\|s-s'\|<\gamma, \\ \|s-s_0\| \geq \delta\}}} \left| Q_{\vec{T}}(s) - Q_{\vec{T}}(s') \right| < c(\gamma) \right\} = 1.$$

For each $\vec{T} \in \mathcal{R}^{m}_{++}(\mathcal{N}_m)$ and $\omega \in \Omega$ the element $s(\vec{T}) = s(\vec{T},\omega)$ is defined by the relation

$$Q_{\vec{T}}\left(s(\vec{T}) \right) = \min_{s \in K} Q_{\vec{T}}(s).$$

Such an element always exists. There may exist more than one minimum point for function $Q_{\vec{T}}$. In this case $s(\vec{T})$ is any minimum point. By Theorem 6 it can be chosen $\mathcal{U}_{\vec{T}}$ – measurable as a function of ω. Then

$$P\left\{\left\|s(\vec{T}) - s_0\right\| \to 0, \quad \vec{T} \to \infty\right\} = 1.$$

If the relationship in condition 4) is

$$P\left\{\overline{\lim_{\vec{T} \to \infty}} \sup_{\|s-s'\|<\gamma} \left|Q_{\vec{T}}(s) - Q_{\vec{T}}(s')\right| < c(\gamma)\right\} = 1$$

then

$$P\left\{Q_{\vec{T}}\left(s(\vec{T})\right) \to \Phi(s_0; s_0), \quad \vec{T} \to \infty\right\} = 1.$$

It is worth to note that the formulation of Theorem 7 differs from that one in [15]. But these differences are unessential and the validity of Theorem 7 is evident.

Lemma 5 *Let $(\Omega, \mathcal{U}, P)$ be a complete probabilistic space and K be a compact subset of some Banach space with a norm $\|\cdot\|$. Assume that $\mathcal{U}_{\vec{T}}$ is a monotone family of σ-algebras, $\mathcal{U}_{\vec{T}_1} \subset \mathcal{U}_{\vec{T}_2}$, $\vec{T}_1 < \vec{T}_2$, (each component of $\vec{T}_1$ is less than the corresponding component of $\vec{T}_2$), $\mathcal{U}_{\vec{T}} \subset \mathcal{U}$, $\vec{T} \in \mathcal{R}_{++}^m(\mathcal{N}^m)$. Suppose that*

$$\left\{Q_{\vec{T}}(s) = Q_{\vec{T}}(s,\omega) : (s,\omega) \in K \times \Omega, \quad \vec{T} \in \mathcal{R}_{++}^m(\mathcal{N}^m)\right\}$$

is the family of real functions satisfying the following conditions:

1) *for fixed $\vec{T}$ and ω the function $Q_{\vec{T}}(s,\omega)$, $s \in K$, is continuous;*

2) *for fixed $\vec{T}$ for each $s \in K$ the function $Q_{\vec{T}}(s,\omega)$ is $\mathcal{U}_{\vec{T}}$-measurable;*

3) *$s_{\vec{T}} = s_{\vec{T}}(\omega)$ is $\mathcal{U}_{\vec{T}}$-measurable family from K and there is the element $s_{\vec{0}} \in K$ such that*

$$P\left\{\lim_{\vec{T} \to \infty} \left\|s_{\vec{T}} - s_{\vec{0}}\right\| = 0\right\} = 1;$$

4) *for any* $s \in K$

$$P\left\{\lim_{\vec{T}\to\infty} Q_{\vec{T}}(s) = \Phi(s)\right\} = 1,$$

where $\Phi(s)$ *is the deterministic continuous function on* K;

5) *there exists* $\gamma_0 > 0$ *such that for all* $0 < \gamma < \gamma_0$ *we have*

$$P\left\{\varlimsup_{\vec{T}\to\infty} \sup_{\|s-s_{\vec{0}}\|<\gamma} \left|Q_{\vec{T}}(s) - Q_{\vec{T}}(s_{\vec{0}})\right| < c(\gamma)\right\} = 1,$$

where $c(\gamma) > 0$ *and* $c(\gamma) \to 0$ *as* $\gamma \to 0$.

Then

$$P\left\{\lim_{\vec{T}\to\infty} Q_{\vec{T}}(s_{\vec{T}}) = \Phi(s_{\vec{0}})\right\} = 1.$$

Proof. The proof is similar to [15], p.80. We will prove the lemma for the case $m = 1$, $\vec{T} = n \in \mathcal{N}$. The general case is proved analogously.

It is sufficient to show that for any $\varepsilon > 0$

$$P\left\{\varlimsup_{n\to\infty} |Q_n(s_n) - \Phi(s_0)| > \varepsilon\right\} = 0.$$

For every $\delta > 0$

$$P\left\{\varlimsup_{n\to\infty} |Q_n(s_n) - \Phi(s_0)| \geq \varepsilon\right\} = P\left\{\omega : \|s_n - s_0\| < \delta, \quad n \geq N(\omega);\right.$$

$$\left.\varlimsup_{n\to\infty} |Q_n(s_n) - Q_n(s_0)| \geq \varepsilon; \quad \varlimsup_{n\to\infty} \sup_{\|s-s_0\|<\gamma} |Q_n(s) - Q_n(s_0)| < c(\gamma)\right\} =$$

$$= P\left\{\varlimsup_{n\to\infty} |Q_n(s_n) - Q_n(s_0)| < c(\delta); \quad \varlimsup_{n\to\infty} |Q_n(s_n) - Q_n(s_0)| \geq \varepsilon\right\}. \quad (2.1)$$

Taking $\delta > 0$ such that $c(\delta) < \varepsilon$, we obtain that the right-hand side of (2.1) vanishes.

2.2 MODELS WITH INDEPENDENT OBSERVATIONS

Let us go to our problems.

Let $\left\{\xi_i,\ i \in \mathcal{N}\right\}$ be a sequence of independent identically distributed random elements defined on a complete probabilistic space $(\Omega, \mathcal{G}, P)$ with values in some metric space $(Y, \mathcal{B}(Y))$. Suppose that I is a closed subset of $\mathcal{R}^l$, $l \geq 1$; probably, $I = \mathcal{R}^l$. Assume that $f : I \times Y \to \mathcal{R}$ is the nonnegative function, satisfying the following conditions:

1) $f(\vec{u}, z)$, $\vec{u} \in I$ is continuous for any $z \in Y$;

2) for each $\vec{u} \in I$ the mapping $f(\vec{u}, z)$, $z \in Y$ is $\mathcal{B}(Y)$ - measurable.

There are observations $\{\xi_i,\ i = \overline{1, n}\}$, $n \in \mathcal{N}$. The problem is to find minimum points of the function

$$F(\vec{u}) = E\left\{ f(\vec{u}, \xi_1) \right\}, \quad \vec{u} \in I, \tag{2.2}$$

and its minimal value.

This task is approximated by searching of minimum points and minimal value for the empirical function

$$F_n(\vec{u}) = F_n(\vec{u}, \omega) = \frac{1}{n} \sum_{i=1}^{n} f(\vec{u}, \xi_i), \quad \vec{u} \in I. \tag{2.3}$$

The following theorem takes place.

Theorem 8 *Suppose that following conditions are satisfied:*

1) for each $c > 0$

$$E\left\{ \max_{\|\vec{u}\| \leq c} f(\vec{u}, \xi_1) \right\} < \infty;$$

2) *for all* $z \in Y'$, $P\{\xi_1 \in Y'\} = 1$, $f(\vec{u}, z) \to \infty$, $\| \vec{u} \| \to \infty$ *(for any $E > 0$
   there exists such a value $\Delta > 0$ that for all $\vec{u}$ with $\| \vec{u} \| > \Delta$ we have
   $f(\vec{u}, z) > E$);*

3) *there exists an unique minimum point $\vec{u}_0$ for the function (2.2).*

Then for each n and $\omega \in \Omega'$, $P(\Omega') = 1$, there exists at least one vector

$$\vec{u}_n = \vec{u}_n(\omega) \in I,$$

*which is the point of the minimum for the function (2.3), and for any n the map-
ping $\vec{u}_n(\omega)$, $\omega \in \Omega'$ can be chosen to be $\mathcal{G}'_n$-measurable, where $\mathcal{G}'_n = \mathcal{G}_n \bigcap \Omega'$,
$\mathcal{G}_n = \sigma\{\xi_i, \ i = \overline{1, n}\}$ is σ-algebra generated by the random elements ξ_i, $i =
\overline{1, n}$. For any choice of the $\mathcal{G}'_n$-measurable function $\vec{u}_n(\omega)$ with probability 1*

$$\vec{u}_n \to \vec{u}_0, \quad F_n(\vec{u}_n) \to F(\vec{u}_0), \quad n \to \infty.$$

Proof. For any n, ω the function $F_n(\vec{u})$, $\vec{u} \in I$ is continuous on I. If I is
unbounded then by condition 2) with probability 1 $F_n(\vec{u}) \to \infty$, $\| \vec{u} \| \to \infty$,
hence there exists such a $\Delta = \Delta_n(\omega) > 0$ that for all $\vec{u} \in I$, $\| \vec{u} \| > \Delta$

$$F_n(\vec{u}) > F_n(\vec{u}_0)$$

and

$$\inf_{\vec{u} \in I} F_n(\vec{u}) = \inf_{\vec{u} \in I \, : \, \| \vec{u} \| \leq \Delta} F_n(\vec{u})$$

with probability 1. If I is bounded then the situation is evident.

For any $\Delta > 0$ the set $\left\{ \vec{u} \in I \, : \, \| \vec{u} \| \leq \Delta \right\}$ is a closed and bounded subset of
$\mathcal{R}^l$, hence it is compact. Then for each n and $\omega \in \Omega'$, $P(\Omega') = 1$ there exists at
least one minimum point $\vec{u}_n = \vec{u}_n(\omega)$ of the function $F_n(\vec{u})$, $\vec{u} \in I$. For all n,
$\vec{u} \in I$ the mapping $F_n(\vec{u}, \omega)$, $\omega \in \Omega'$ is $\mathcal{G}'_n$-measurable. By Theorem 6 for any
n the mapping $\vec{u}_n(\omega)$, $\omega \in \Omega'$ can be chosen to be $\mathcal{G}'_n$-measurable.

It will be shown that if I is unbounded then it can be found such a $c > 0$
that beginning from some n, which depends on ω, all minimum points of the
function $F_n(\vec{u})$, $\vec{u} \in I$ belong to the set

$$K_c = \left\{ \vec{u} \in I \, : \, \| \vec{u} \| \leq c \right\}$$

with probability 1.

For any $c > 0$ the mapping

$$\varphi(c, z) = \inf_{\|\vec{u}\| > c} f(\vec{u}, z), \quad z \in Y$$

is $\mathcal{B}(Y)$-measurable. Because of condition 2) for any i

$$\varphi(c, \xi_i) \to \infty, \quad c \to \infty$$

with probability 1.

Then by Lemma 4

$$E\left\{\varphi(c, \xi_1)\right\} \to \infty, \quad c \to \infty.$$

It can be chosen such a c_0 that

$$E\left\{\varphi(c_0, \xi_1)\right\} > E\left\{f(\vec{u}_0, \xi_1)\right\}.$$

By Theorem 1 beginning from some n, which depends on ω,

$$\frac{1}{n}\sum_{i=1}^{n}\varphi(c_0, \xi_i) > \frac{1}{n}\sum_{i=1}^{n}f(\vec{u}_0, \xi_i) = F_n(\vec{u}_0) \qquad (2.4)$$

with probability 1. Since

$$\inf_{\|\vec{u}\| > c_0} F_n(\vec{u}) \geq \frac{1}{n}\sum_{i=1}^{n}\varphi(c_0, \xi_i)$$

(2.4) implies the fact we need.

Since $\mathcal{R}^l$ is a Banach space, then a closed bounded subset of $\mathcal{R}^l$ is a compact. Hence to prove Theorem 8 it is sufficient to check conditions of Theorem 7 for the sequence of functions

$$\left\{F_n : K \times \Omega' \to \mathcal{R}, \quad n \geq 1\right\}, \qquad (2.5)$$

where $K = K_{c_0}$.

It is evident that conditions 1) and 2) of Theorem 7 are fulfilled. By Theorem 1

$$P\left\{\lim_{n\to\infty} F_n(\vec{u}) = F(\vec{u})\right\} = 1, \quad \vec{u} \in I.$$

The condition 1) and Lebesgue theorem of limit transition imply that the function $F(\vec{u})$, $\vec{u} \in I$ is continuous. Then the condition 3) of Theorem 7 is satisfied.

Denote

$$\psi(\gamma, z) = \sup_{\vec{u}, \vec{v} \in K : \|\vec{u} - \vec{v}\| < \gamma} \left| f(\vec{u}, z) - f(\vec{v}, z) \right|; \quad \gamma > 0, \quad z \in Y.$$

By Lemma 3 for any $\gamma > 0$ the mapping $\psi(\gamma, z)$, $z \in Y$ is $\mathcal{B}(Y)$-measurable. For each n, $\vec{u}_1 \in K$, $\gamma > 0$

$$\zeta_n(\vec{u}_1, \gamma) = \sup_{\vec{u} \in K : \|\vec{u} - \vec{u}_1\| < \gamma} \left| F_n(\vec{u}) - F_n(\vec{u}_1) \right| \leq \frac{1}{n} \sum_{i=1}^{n} \psi(\gamma, \xi_i) \qquad (2.6)$$

with probability 1. Theorem 1 implies that

$$P \left\{ \lim_{n \to \infty} \frac{1}{n} \sum_{i=1}^{n} \psi(\gamma, \xi_i) = E \left\{ \psi(\gamma, \xi_1) \right\} \right\} = 1, \quad \gamma > 0. \qquad (2.7)$$

Denote

$$c(\gamma) = E \left\{ \psi(\gamma, \xi_1) \right\} + \gamma, \quad \gamma > 0.$$

From (2.6), (2.7) for all $\vec{u}_1 \in K$, $\gamma > 0$ we have

$$P \left\{ \overline{\lim_{n \to \infty}} \, \zeta_n(\vec{u}_1, \gamma) < c(\gamma) \right\} = 1.$$

If the function is continuous on a compact then it is uniformly continuous. Hence for any $z \in Y$

$$\psi(\gamma, z) \searrow 0, \quad \gamma \to 0.$$

Then by B.Levi theorem of limit transition under an integral sign for a monotone sequence of functions

$$c(\gamma) \to 0, \quad \gamma \to 0.$$

For the sequence of functions (2.5) all conditions of Theorem 7 are fulfilled. The theorem is proved.

Theorem 9 *Suppose that*

$$E \left\{ f(\vec{u}, \xi_1) \right\} < \infty, \quad \vec{u} \in I.$$

Let $\vec{u}_0$ and $\vec{u}_n = \vec{u}_n(\omega)$ be some minimum points of functions (2.2) and (2.3) respectively; $n \geq 1$, $\omega \in \Omega'$, $P(\Omega') = 1$ (it is supposed that functions (2.2) and (2.3) have minimum points); for any n the mapping $\vec{u}_n(\omega)$, $\omega \in \Omega'$ is supposed to be $\mathcal{G}'_n$-measurable. Let $\vec{u}_n \rightarrow \vec{u}_0$, $n \rightarrow \infty$ with probability 1 and the following conditions be fulfilled:

1) $\vec{u}_0$ is an internal point of I;

2) there exists such a closed neighbourhood S of $\vec{u}_0$ that for any $z \in Y$ the function $f(\vec{u}, z)$, $\vec{u} \in S$ is twice continuously differentiable on S;

3) $E \left\{ \max_{\vec{u} \in S} \left\| \vec{\nabla} f(\vec{u}, \xi_1) \right\| \right\} < \infty$, where

$$\vec{\nabla} f(\vec{u}, z) = \left(\frac{\partial f}{\partial u_j}(\vec{u}, z) \right)_{j=1}^{l}, \quad \vec{u} = (u_j)_{j=1}^{l} \in S, \quad z \in Y,$$

(due to properties of measurable functions for all $\vec{u} \in S$ the mapping $\vec{\nabla} f(\vec{u}, z)$, $z \in Y$ is $\mathcal{B}(Y)$-measurable);

4) $E \left\{ \max_{\vec{u} \in S} \left\| \Phi(\vec{u}, \xi_1) \right\| \right\} < \infty$, where

$$\Phi(\vec{u}, z) = \left(\frac{\partial^2 f}{\partial u_j \, \partial u_k}(\vec{u}, z) \right)_{j,k=1}^{l};$$

5) $\det A_0 \neq 0$, where

$$A_0 = E \left\{ \Phi(\vec{u}_0, \xi_1) \right\};$$

6) $E \left\{ \left\| \vec{\nabla} f(\vec{u}_0, \xi_1) \right\|^2 \right\} < \infty$, and $\det C \neq 0$, where

$$C = E \left\{ \vec{\nabla} f(\vec{u}_0, \xi_1) \left(\vec{\nabla} f(\vec{u}_0, \xi_1) \right)' \right\}.$$

Then $\sqrt{n}(\vec{u}_n - \vec{u}_0)$ converges in distribution to $\mathcal{N}(\vec{0}, (A_0)^{-1} C (A_0)^{-1})$, $n \rightarrow \infty$, where $\mathcal{N}(\vec{a}, B)$ is Gaussian distribution with the mean $\vec{a}$ and the correlation matrix B.

Let following conditions also be satisfied:

7) $E\left\{\left(f(\vec{u}_0,\xi_1)\right)^2\right\} < \infty;$

8) $\sigma^2 = E\left\{\left(f(\vec{u}_0,\xi_1) - E\{f(\vec{u}_0,\xi_1)\}\right)^2\right\} > 0.$

Then $\sqrt{n}\left(F_n(\vec{u}_n) - F(\vec{u}_0)\right)$ converges in distribution to $\mathcal{N}(0,\sigma^2)$, $n \to \infty$.

Proof. By condition 2) the function $F_n(\vec{u})$ is differentiable on S and

$$\vec{\nabla} F_n(\vec{u}) = \left(\frac{\partial F_n}{\partial u_j}(\vec{u})\right)^l_{j=1} = \frac{1}{n}\sum_{i=1}^{n}\vec{\nabla} f(\vec{u},\xi_i), \quad \vec{u}\in S.$$

Since $\vec{u}_n$ is the consistent estimate of $\vec{u}_0$ then it is an internal point of S with probability, converging to $1, n \to \infty$. Since $\vec{u}_n$ is a minimum point of F_n then with the same probability

$$\vec{\nabla} F_n(\vec{u}_n) = \vec{0}.$$

Under condition 2) the function $F_n(\vec{u})$, $\vec{u}\in S$ is twice continuously differentiable on S and

$$\frac{\partial^2 F_n}{\partial u_j\,\partial u_k}(\vec{u}) = \frac{1}{n}\sum_{i=1}^{n}\frac{\partial^2 f}{\partial u_j\,\partial u_k}(\vec{u},\xi_i); \quad j,k = \overline{1,l}, \quad \vec{u}\in S.$$

For fixed n,ω Tailor formula [101] can be applied to each component of the vector-function $\vec{\nabla} F_n(\vec{u})$, $\vec{u}\in S$. Then for any $n,\omega \in \Omega_n$, $P(\Omega_n) \to 1$, $n \to \infty$

$$\vec{\nabla} F_n(\vec{u}_0) + A_n(\vec{u}_n - \vec{u}_0) = \vec{0}, \tag{2.8}$$

where

- $A_n = A_n(\omega) = \left(\dfrac{\partial^2 F_n}{\partial u_j\,\partial u_k}\left(\vec{u}^j(n,\omega),\omega\right)\right)^l_{j,k=1};$

- $\vec{u}^j(n,\omega) = \vec{u}_0 + \theta_j(n,\omega)(\vec{u}_n(\omega) - \vec{u}_0);$

- $\theta_j(n,\omega)$ is a minimum point of the function $\varphi_j(n,t,\omega)$, $t \in [0,1]$, which is $\mathcal{G}$-measurable in ω for any n;

-

$$
\begin{aligned}
\varphi_j(n,t,\omega) \;=\; & \psi_j(n,t,\omega) \; \chi \left(\bigvee_{\tau \in]0,1[\bigcap Q} \{\psi_j(n,\tau,\omega) < \psi_j(n,0,\omega)\} \right) - \\[2mm]
& - \psi_j(n,t,\omega) \; \chi \left(\bigwedge_{\tau \in]0,1[\bigcap Q} \{\psi_j(n,\tau,\omega) \geq \psi_j(n,0,\omega)\} \right) \bigwedge \\[2mm]
& \bigwedge \left(\bigvee_{\tau \in]0,1[\bigcap Q} \{\psi_j(n,\tau,\omega) > \psi_j(n,0,\omega)\} \right) + \\[2mm]
& + \; |1/2 - t| \; \chi \left(\bigwedge_{\tau \in]0,1[\bigcap Q} \{\psi_j(n,\tau,\omega) = \psi_j(n,0,\omega)\} \right) ;
\end{aligned}
$$

- $\psi_j(n,\tau,\omega) = \dfrac{\partial F_n}{\partial u_j}\left(\vec{u}_0 + t(\vec{u}_n - \vec{u}_0), \omega\right) + t\dfrac{\partial F_n}{\partial u_j}(\vec{u}_0, \omega); t \in [0,1], j = \overline{1,l}$;

- $\chi(C) = \begin{cases} 1, & \text{if the condition C is true} \\ 0, & \text{if it is false;} \end{cases}$

- $\bigvee\limits_{i \in I} A_i$ is a condition which is true if and only if at least one of conditions A_i is true;

- $\bigwedge\limits_{i \in I} A_i$ is a condition which is true if and only if all of conditions A_i are true.

By Theorem 6 and Lemma 1 for any n the mappings $\theta_j(n,\omega)$, $\omega \in \Omega_n$, $j = \overline{1,l}$ can be chosen $\mathcal{G}$-measurable. Lemma 1 implies $\mathcal{G}$-measurability of the function $A_n(\omega)$, $\omega \in \Omega_n$ for all n.

Denote

$$A_1(n) = A_1(n,\omega) = \begin{cases} A_n(\omega), & \omega \in \Omega_n, \\ \\ A_0, & \omega \in \Omega \setminus \Omega_n, \end{cases} \quad n \geq 1, \quad \omega \in \Omega,$$

$$\Psi_n(\vec{u}) = \Psi_n(\vec{u},\omega) = \begin{cases} \left(\dfrac{1}{n} \displaystyle\sum_{i=1}^{n} \dfrac{\partial^2 f}{\partial u_j\, \partial u_k}\left(\vec{u},\xi_i(\omega)\right) \right)^l_{j,k=1}, & \omega \in \Omega_n, \\ \\ \Psi(\vec{u}) = E\left\{ \Phi(\vec{u},\xi_1) \right\}, & \omega \in \Omega \setminus \Omega_n, \end{cases}$$

$\vec{u} \in S.$

Let us apply Lemma 5 to the sequence of functions $\Psi_n^{jk}(\vec{u})$, where

$$\Psi_n(\vec{u}) = \left(\Psi_n^{jk}(\vec{u}) \right)^l_{j,k=1}.$$

Fix j,k. By the strong law of large numbers and condition 4)

$$P\left\{ \lim_{n\to\infty} \Psi_n^{jk}(\vec{u}) = \Psi^{jk}(\vec{u}) \right\} = 1,$$

where $\Psi(\vec{u}) = \left(\Psi^{jk}(\vec{u}) \right)^l_{j,k=1}$. Then we have

$$\sup_{\|\vec{u}-\vec{u}_0\|<\gamma} \left| \Psi_n^{jk}(\vec{u}) - \Psi_n^{jk}(\vec{u}_0) \right| \leq \frac{1}{n} \sum_{i=1}^{n} \sup_{\|\vec{u}-\vec{u}_0\|<\gamma} \left| \frac{\partial^2 f}{\partial u_j\, \partial u_k}(\vec{u},\xi_i) - \right.$$

$$\left. - \frac{\partial^2 f}{\partial u_j\, \partial u_k}(\vec{u}_0,\xi_i) \right| = \zeta_n(\gamma), \quad \omega \in \Omega_n,$$

$$\sup_{\|\vec{u}-\vec{u}_0\|<\gamma} \left| \Psi_n^{jk}(\vec{u}) - \Psi_n^{jk}(\vec{u}_0) \right| \leq E \sup_{\|\vec{u}-\vec{u}_0\|<\gamma} \left| \frac{\partial^2 f}{\partial u_j\, \partial u_k}(\vec{u},\xi_1) - \right.$$

$$\left. - \frac{\partial^2 f}{\partial u_j\, \partial u_k}(\vec{u}_0,\xi_1) \right| = c(\gamma) - \gamma, \quad \omega \in \Omega \setminus \Omega_n.$$

By virtue of the strong law of large numbers

$$P\left\{\,\zeta_n(\gamma) \to c(\gamma) - \gamma, \quad n \to \infty\right\} = 1.$$

Then

$$P\left\{\varlimsup_{n\to\infty} \sup_{\|\vec{u}-\vec{u}_0\|<\gamma} \left|\Psi_n^{jk}(\vec{u}) - \Psi_n^{jk}(\vec{u}_0)\right| < c(\gamma)\right\} = 1.$$

Since $\Phi(\vec{u}, z)$ is continuous in the first argument, using B.Levi theorem of passing to a limit under an integral symbol we have $c(\gamma) \to 0$, $\gamma \to 0$.

Denote

$$\vec{s}_n = \vec{s}_n(\omega) = \begin{cases} \vec{u}^j(n,\omega), & \omega \in \Omega_n, \\ \vec{u}_0, & \omega \in \Omega \setminus \Omega_n, \end{cases} \quad , \quad n \geq 1.$$

Then

$$P\left\{\vec{s}_n \to \vec{u}_0, \quad n \to \infty\right\} = 1.$$

All conditions of Lemma 5 are fulfilled. Thus,

$$P\left\{\Psi_n^{jk}(\vec{s}_n) \to \Psi^{jk}(\vec{u}_0), \quad n \to \infty\right\} = 1.$$

Consequently,

$$P\left\{A_1(n) \to A_0, \quad n \to \infty\right\} = 1. \tag{2.9}$$

Since $\det A_0 \neq 0$ then $\det A_1(n) \neq 0$ for any n, $\omega \in \Omega_1(n)$, $P\left(\Omega_1(n)\right) \to 1$, $n \to \infty$. We obtain from (2.8) that

$$\sqrt{n}\,(\vec{u}_n - \vec{u}_0) = -\left(A_1(n)\right)^{-1} \frac{1}{\sqrt{n}} \sum_{i=1}^{n} \vec{\nabla} f(\vec{u}_0, \xi_i). \tag{2.10}$$

Denote

$$A_2(n) = A_2(n,\omega) = \begin{cases} \left(A_1(n,\omega)\right)^{-1}, & \omega \in \Omega_1(n), \\ (A_0)^{-1}, & \omega \in \Omega \setminus \Omega_1(n), \end{cases} \quad n \geq 1, \quad \omega \in \Omega.$$

It follows from (2.9) that

$$P\Big\{ A_2(n) \to (A_0)^{-1}, \quad n \to \infty \Big\} = 1.$$

Conditions 2), 3) and Lebesgue theorem of limit transition imply that

$$\vec{\nabla} E\Big\{ f(\vec{u}, \xi_1) \Big\} = \left(\frac{\partial E\Big\{ f(\vec{u}, \xi_1) \Big\}}{\partial u_j} \right)^l_{j=1} = E\Big\{ \vec{\nabla} f(\vec{u}, \xi_1) \Big\}, \quad \vec{u} \in S.$$

Hence

$$E\Big\{ \vec{\nabla} f(\vec{u}_0, \xi_1) \Big\} = \vec{0} .$$

Because of condition 6) and Theorem 3

$$\frac{1}{\sqrt{n}} \sum_{i=1}^n \vec{\nabla} f(\vec{u}_0, \xi_i)$$

converges in distribution to $\mathcal{N}(\vec{0}, C)$, $n \to \infty$.

Then we have that (2.10), Lemma 2 and probability properties imply the validity of the first part of Theorem 9.

By Tailor formula for all n, $\omega \in \Omega_n$, $P(\Omega_n) \to 1$, $n \to \infty$, we have

$$F_n(\vec{u}_n) - F_n(\vec{u}_0) = \Big(\vec{\nabla} F_n(\vec{u}_0) \Big)' (\vec{u}_n - \vec{u}_0) +$$

$$+1/2\,(\vec{u}_n - \vec{u}_0)' \left(\frac{\partial^2 F_n}{\partial u_j \partial u_k}(\vec{v}_n) \right)^l_{j,k=1} (\vec{u}_n - \vec{u}_0),$$

where

- $\vec{v}_n = \vec{v}_n(\omega) = \vec{u}_0 + \theta_n(\omega)\Big(\vec{u}_n(\omega) - \vec{u}_0 \Big);$

- $\theta_n(\omega)$ is an extremum point of the function $h_n(t, \omega)$, $t \in\,]0, 1[;$

- $$h_n(t,\omega) = F_n\left(\vec{u}_0 + t\left(\vec{u}_n(\omega) - \vec{u}_0\right), w\right) + (1-t) \times$$
$$\times \left(\vec{\nabla} F_n(\vec{u}_0 + t(\vec{u}_n(\omega) - \vec{u}_0), w)\right) \times (\vec{u}_n(\omega) - \vec{u}_0) +$$
$$+ (1-t^2)\left(F_n(\vec{u}_n(\omega), \omega) - F_n(\vec{u}_0, \omega) -\right.$$
$$- \left.\left(\vec{\nabla} F_n(\vec{u}_0, \omega)\right)'(\vec{u}_n - \vec{u}_0);\right.$$

- for each n the mapping $\theta_n(\omega)$, $\omega \in \Omega_n$ is $\mathcal{G}$-measurable.

Denote

$$A_3(n) = A_3(n, \omega) = \begin{cases} \left(\dfrac{\partial^2 F_n}{\partial u_j \partial u_k}(\vec{v}_n(\omega), \omega)\right)^l_{j,k=1}, & \omega \in \Omega_n, \\[2mm] A_0, & \omega \in \Omega \setminus \Omega_n \end{cases},$$

$$n \geq 1, \quad \omega \in \Omega.$$

From (2.9) it follows that

$$P\left\{A_3(n) \to A_0, \quad n \to \infty\right\} = 1.$$

For any n, $\omega \in \Omega_n$

$$\sqrt{n}\left(F_n(\vec{u}_n) - F(\vec{u}_0)\right) = \left(\frac{1}{\sqrt{n}}\sum_{i=1}^n \vec{\nabla} f(\vec{u}_0, \xi_i)\right)'(\vec{u}_n - \vec{u}_0) +$$

$$+ \frac{\sqrt{n}}{2}(\vec{u}_n - \vec{u}_0)' A_3(n)(\vec{u}_n - \vec{u}_0) +$$

$$+ \frac{1}{\sqrt{n}}\sum_{i=1}^n \left(f(\vec{u}_0, \xi_i) - E\left\{f(\vec{u}_0, \xi_1)\right\}\right).$$

$$(2.11)$$

If a sequence of random variables converges in distribution to a constant, then it converges in probability to this constant. Then by Lemma 2

$$\left(\frac{1}{\sqrt{n}}\sum_{i=1}^n \vec{\nabla} f(\vec{u}_0, \xi_i)\right)'(\vec{u}_n - \vec{u}_0) \to 0,$$

$$\frac{\sqrt{n}}{2} \left(\vec{u}_n - \vec{u}_0 \right)' A_3(n) \left(\vec{u}_n - \vec{u}_0 \right) \to 0, \quad n \to \infty$$

in probability. Conditions 7), 8) imply that

$$\frac{1}{\sqrt{n}} \sum_{i=1}^{n} \left(f(\vec{u}_0, \xi_i) - E\left\{ f(\vec{u}_0, \xi_1) \right\} \right) \Longrightarrow \mathcal{N}(0, \sigma^2), \quad n \to \infty,$$

where "$\Longrightarrow$" means the convergence in distribution. Then Lemma 2 implies the validity of the second part of Theorem 9.

Let us consider the more comlicated problem.

Suppose that $\left\{ \xi_i, i \in Z \right\}$ is a stationary in a strict sense ergodic random process with a discrete parameter, defined on a complete probabilistic space $(\Omega, \mathcal{G}, P)$, with values in some metric space $\left(Y, \mathcal{B}(Y) \right)$; I is a closed subset of $\mathcal{R}^l$, $l \geq 1$; $f : I \times Y \to \mathcal{R}$ is a nonnegative function, continuous in the first argument and measurable in the second one.

We have the observations

$$\left\{ \xi_i, i = \overline{1, n} \right\}, \quad n \geq 1.$$

The task is to find minimum points and the minimal value of the function

$$F(\vec{u}) = E\left\{ f(\vec{u}, \xi_1) \right\}, \quad \vec{u} \in I. \tag{2.12}$$

Let us define an empirical function

$$F_n(\vec{u}) = F_n(\vec{u}, \omega) = \frac{1}{n} \sum_{i=1}^{n} f(\vec{u}, \xi_i), \quad \vec{u} \in I. \tag{2.13}$$

The following theorem takes place.

Theorem 10 *Suppose that the following conditions are satisfied:*

1) for all $c > 0$

$$E\left\{ \max_{\|\vec{u}\| \leq c} f(\vec{u}, \xi_1) \right\} < \infty;$$

2) for any $z \in Y'$, $P\{\xi_1 \in Y'\} = 1$, $f(\vec{u}, z) \to \infty$, $\| \vec{u} \| \to \infty$;

3) there exists an unique minimum point $\vec{u}_0$ of the function (2.12).

Then for any n and $\omega \in \Omega'$, $P(\Omega') = 1$, there exists at least one minimum point $\vec{u}_n = \vec{u}_n(\omega)$ of the function (2.13), and for each n the mapping $\vec{u}_n(\omega)$, $\omega \in \Omega'$, can be chosen to be $\mathcal{G}'_n$-measurable, $\mathcal{G}'_n = \mathcal{G}_n \cap \Omega'$, $\mathcal{G}_n = \sigma\{\xi_i, i = \overline{1,n}\}$. For any choice of the $\mathcal{G}'_n$-measurable function $\vec{u}_n(\omega)$

$$P\left\{\vec{u}_n \to \vec{u}_0, \quad F_n(\vec{u}_n) \to F(\vec{u}_0), \quad n \to \infty\right\} = 1.$$

Proof. For any n, ω the function $F_n(\vec{u})$, $\vec{u} \in I$ is continuous. By condition 2) for any n we have

$$P\left\{F_n(\vec{u}) \to \infty, \quad \| \vec{u} \| \to \infty\right\} = 1,$$

hence for each n there exists such $\Delta = \Delta(n, \omega)$ that for any $\vec{u}$ with $\| \vec{u} \| > \Delta$

$$F_n(\vec{u}) > F_n(\vec{u}_0)$$

with probability 1. Then for such Δ

$$\inf_{\vec{u} \in I} F_n(\vec{u}) = \inf_{\vec{u} \in I : \|\vec{u}\| \leq \Delta} F_n(\vec{u})$$

with probability 1.

The set $\left\{\vec{u} \in I : \| \vec{u} \| \leq \Delta\right\}$ is compact which implies the existance of a minimum point for the function (2.13). By Theorem 6 for each n the mapping $\vec{u}_n(\omega)$, $\omega \in \Omega'$, can be chosen $\mathcal{G}'_n$-measurable.

As in Theorem 8 it can be proved that there exists such $c > 0$ that, beginning from some n, which depends on ω, all minimum points of the function $F_n(\vec{u})$ belong to the set

$$K_c = \left\{\vec{u} \in I : \| \vec{u} \| \leq c\right\}$$

with probability 1. But in this proof we use the properties of ergodic random processes instead of Theorem 1.

Then to prove Theorem 10 it is sufficient to check conditions of Theorem 7 for the sequence of functions

$$\left\{ F_n \ : \ K \times \Omega' \to \mathcal{R}, \quad n \geq 1 \right\}, \tag{2.14}$$

where $K = K_c$.

The conditions 1) and 2) are fulfilled. By properties of ergodic random processes

$$P\left\{ \lim_{n \to \infty} F_n(\vec{u}) = F(\vec{u}) \right\} = 1, \quad \vec{u} \in I.$$

In view of condition 1) of Theorem 10 and Lebesgue theorem of limit transition the function $F(\vec{u})$ is continuous. Then condition 3) of Theorem 7 is satisfied.

The validity of the fourth condition of Theorem 7 is checked by the same way as in Theorem 8. The proof is complete.

Lemma 6 *[54] Suppose that $\left\{ \vec{\eta}_i, \ i \in Z \right\}$ is measurable stationary in a strict sense random process, defined on a complete probabilistic space $(\Omega, \mathcal{G}, P)$, with values in $\mathcal{R}^m$, $m \geq 1$.*

1. *Let the process $\vec{\eta}_i$ satisfy a strong mixing condition with the coefficient $\alpha(k) = O(k^{-1-\varepsilon})$, $k \to \infty$; $\varepsilon > 0$.*

2. *Assume that*

$$E\left\{ \| \vec{\eta}_1 \|^{2+\delta} \right\} < \infty; \quad \delta > 4/\varepsilon.$$

Then the process $\vec{\eta}_i$ has a spectral density, which is bounded and continuous.

Theorem 11 *Suppose that*

$$E\left\{ f(\vec{u}, \xi_1) \right\} < \infty, \quad \vec{u} \in I.$$

Let $\vec{u}_0$ and $\vec{u}_n = \vec{u}_n(\omega)$ be minimum points of the functions (2.12) and (2.13) respectively; $n \geq 1$, $\omega \in \Omega'$, $P(\Omega') = 1$. Assume that for any n the mapping $\vec{u}_n(\omega)$, $\omega \in \Omega'$ is $\mathcal{G}'_n$-measurable. Let

$$P\left\{ \vec{u}_n \to \vec{u}_0, \quad n \to \infty \right\} = 1,$$

conditions 1) − 5) of Theorem 9 and the following conditions

6) *the process ξ_i satisfies a strong mixing condition with the coefficient $\alpha(k) = O(k^{-1-\varepsilon})$, $k \to \infty$; $\varepsilon > 0$;*

7) *for some $\delta > 4/\varepsilon$*

$$E\left\{\left\|\vec{\nabla} f(\vec{u}_0, \xi_1)\right\|^{2+\delta}\right\} < \infty;$$

8) $\det g(0) \neq 0$, *where* $g(\lambda) = \left(g_{jk}(\lambda)\right)_{j,k=1}^{l}$ *is the spectral density of the process* $\left\{\vec{\nabla} f(\vec{u}_0, \xi_i), \, i \in Z\right\}$

be fulfilled.

Then $\sqrt{n}\,(\vec{u}_n - \vec{u}_0)$ converges in distribution to $\mathcal{N}\left(\vec{0}, 2\pi\,(A_0)^{-1} g(0)\,(A_0)^{-1}\right)$, $n \to \infty$.

Suppose that also the following conditions are satisfied:

9) *for some $\delta_1 > 4/\varepsilon$*

$$E\left\{\left(f(\vec{u}_0, \xi_1)\right)^{2+\delta_1}\right\} < \infty;$$

10) $g_1(0) \neq 0$, *where $g_1(\lambda)$ is the spectral density of the process* $\left\{f(\vec{u}_0, \xi_i), \, i \in Z\right\}$.

Then $\sqrt{n}\left(F_n(\vec{u}_n) - F(\vec{u}_0)\right)$ converges in distribution to $\mathcal{N}\left(0, 2\pi g_1(0)\right)$, $n \to \infty$.

Proof. As in Theorem 9 for any n, $\omega \in \Omega_n$, $P(\Omega_n) \to 1$, $n \to \infty$, we may write

$$\vec{\nabla} F_n(\vec{u}_0) + A_n(\vec{u}_n - \vec{u}_0) = \vec{0}, \tag{2.15}$$

where

$$A_n = A_n(\omega) = \left(\frac{\partial^2 F_n}{\partial u_j \partial u_k}\left(\vec{u}^j(n, \omega), \omega\right)\right)_{j,k=1}^{l}.$$

Denote

$$A_1(n) = A_1(n,\omega) = \begin{cases} A_n(\omega), & \omega \in \Omega_n \\ \\ A_0, & \omega \in \Omega \setminus \Omega_n \end{cases}, \quad n \geq 1, \quad \omega \in \Omega.$$

Analogously to Theorem 9

$$P\left\{A_1(n) \to A_0, \quad n \to \infty\right\} = 1.$$

As in Theorem 9 for any n, $\omega \in \Omega_1(n)$, $P\left(\Omega_1(n)\right) \to 1$, $n \to \infty$, $\det\left(A_1(n)\right) \neq 0$ and (2.15) implies that

$$\sqrt{n}\left(\vec{u}_n - \vec{u}_0\right) = -\left(A_1(n)\right)^{-1} \frac{1}{\sqrt{n}} \sum_{i=1}^{n} \vec{\nabla} f(\vec{u}_0, \xi_i). \qquad (2.16)$$

Denote

$$A_2(n) = A_2(n,\omega) = \begin{cases} (A_1(n,\omega))^{-1}, & \omega \in \Omega_1(n) \\ \\ (A_0)^{-1}, & \omega \in \Omega \setminus \Omega_1(n) \end{cases},$$

$$n \geq 1, \quad \omega \in \Omega.$$

Then

$$P\left\{A_2(n) \to (A_0)^{-1}, \quad n \to \infty\right\} = 1.$$

As in Theorem 9

$$E\left\{\vec{\nabla} f(\vec{u}_0, \xi_1)\right\} = \vec{0}.$$

By conditions 6) - 8) and Theorem 4

$$\frac{1}{\sqrt{n}} \sum_{i=1}^{n} \vec{\nabla} f(\vec{u}_0, \xi_i)$$

converges in distribution to $\mathcal{N}\left(\vec{0}, 2\pi g(0)\right)$, $n \to \infty$. Probability properties and (2.16) imply the validity of the first part of Theorem 11.

By Tailor formula for any n, $\omega \in \Omega_n$, $P(\Omega_n) \to 1$, $n \to \infty$

$$F_n(\vec{u}_n) - F_n(\vec{u}_0) = \left(\vec{\nabla} F_n(\vec{u}_0)\right)' (\vec{u}_n - \vec{u}_0) +$$

$$+ \frac{1}{2}(\vec{u}_n - \vec{u}_0)' \left(\frac{\partial^2 F_n}{\partial u_j \partial u_k}(\vec{v}_n)\right)^l_{j,k=1} (\vec{u}_n - \vec{u}_0).$$

Denote

$$A_3(n) = A_3(n, \omega) = \begin{cases} \left(\dfrac{\partial^2 F_n}{\partial u_j \partial u_k}(\vec{v}_n(\omega), \omega)\right)^l_{j,k=1}, & \omega \in \Omega_n \\ A_0, & \omega \in \Omega \setminus \Omega_n \end{cases},$$

$$n \geq 1, \quad \omega \in \Omega.$$

Then

$$P\left\{A_3(n) \to A_0, \quad n \to \infty\right\} = 1.$$

For any n, $\omega \in \Omega_n$

$$\sqrt{n}\left(F_n(\vec{u}_n) - F(\vec{u}_0)\right) = \left(\frac{1}{\sqrt{n}} \sum_{i=1}^{n} \vec{\nabla} f(\vec{u}_0, \xi_i)\right)' (\vec{u}_n - \vec{u}_0) +$$

$$+ \frac{\sqrt{n}}{2}(\vec{u}_n - \vec{u}_0)' A_3(n)(\vec{u}_n - \vec{u}_0) +$$

$$+ \frac{1}{\sqrt{n}} \sum_{i=1}^{n} \left(f(\vec{u}_0, \xi_i) - E\left\{f(\vec{u}_0, \xi_1)\right\}\right).$$

$$(2.17)$$

As in Theorem 9

$$\left(\frac{1}{\sqrt{n}} \sum_{i=1}^{n} \vec{\nabla} f(\vec{u}_0, \xi_i)\right)' (\vec{u}_n - \vec{u}_0) \to 0,$$

$$\frac{\sqrt{n}}{2}(\vec{u}_n - \vec{u}_0)' A_3(n)(\vec{u}_n - \vec{u}_0) \to 0, \quad n \to \infty$$

in probability. By conditions 6), 9), 10) and Theorem 4

$$\frac{1}{\sqrt{n}} \sum_{i=1}^{n} \left(f(\vec{u}_0, \xi_i) - E\left\{ f(\vec{u}_0, \xi_1) \right\} \right)$$

converges in distribution to $\mathcal{N}\left(0, 2\pi g_1(0)\right)$, $n \to \infty$. Lemma 2 implies that the second part of Theorem 11 is true.

Now we will consider the case when the time is continuous .

2.3   MODELS WITH CONTINUOUS TIME

Suppose that $\left\{ \xi(t) = \xi(t, \omega), \ t \in \mathcal{R} \right\}$ is a measurable stationary in a strict sense ergodic random process, defined on a complete probabilistic space $(\Omega, \mathcal{G}, P)$, with values in some metric space $\left(Y, \mathcal{B}(Y) \right)$; I is a closed subset of $\mathcal{R}^l$, $l \geq 1$, perhaps, $I = \mathcal{R}^l$; $f : I \times Y \to \mathcal{R}$ is a nonnegative function satisfying the following conditions:

1) the function $f(\vec{u}, z)$, $\vec{u} \in I$ is continuous for any $z \in Y$;

2) for each $\vec{u} \in I$ the mapping $f(\vec{u}, z)$, $z \in Y$ is $\mathcal{B}(Y)$-measurable.

Consider the observations

$$\left\{ \xi(t), \ t \in [0, T] \right\}, \quad T > 0.$$

The task is to find vectors from I, which are minimum points for the function

$$F(\vec{u}) = E\left\{ f\left(\vec{u}, \xi(0) \right) \right\}, \quad \vec{u} \in I \tag{2.18}$$

and the minimal value of this function.

The problem is approximated by the task of searching of minimum points and the minimal value for the empirical function

$$F_T(\vec{u}) = F_T(\vec{u}, \omega) = \frac{1}{T} \int_0^T f\left(\vec{u}, \xi(t) \right) dt, \quad \vec{u} \in I, \tag{2.19}$$

where the integral is taken in Lebesgue sense.

The following theorem takes place.

Theorem 12 *Assume that the following conditions are satisfied:*

1) for all $c > 0$

$$E\left\{ \max_{\|\vec{u}\| \le c} f\left(\vec{u}, \xi(0)\right) \right\} < \infty;$$

2) if I is unbounded then for any $z \in Y'$, $P\left\{\xi(t) \in Y' \text{ for all } t \ge 0\right\} = 1$, we have

$$f(\vec{u}, z) \to \infty, \quad \|\vec{u}\| \to \infty;$$

3) there exists the unique vector $\vec{u}_0 \in I$, which is a minimum point of the function (2.18).

Then for all $T > 0$ and $\omega \in \Omega'$, $P(\Omega') = 1$ there exists at least one vector $\vec{u}(T) = \vec{u}(T, \omega) \in I$, which is a minimum point of the function (2.19), and for each $T > 0$ the mapping $\vec{u}(T, \omega)$, $\omega \in \Omega'$ can be chosen $\mathcal{G}'_T$-measurable, where $\mathcal{G}'_T = \mathcal{G}_T \cap \Omega'$, $\mathcal{G}_T = \sigma\left\{\xi(t), \ 0 \le t \le T\right\}$. For any choice of the $\mathcal{G}'_T$-measurable function $\vec{u}(T, \omega)$

$$P\left\{\vec{u}(T) \to \vec{u}_0, \quad F_T\left(\vec{u}(T)\right) \to F(\vec{u}_0), \quad T \to \infty\right\} = 1.$$

Proof. By condition 1) and Fubini theorem for any n

$$P\left\{\int_0^n \max_{\|\vec{u}\| \le n} f\left(\vec{u}, \xi(t)\right) dt < \infty\right\} = 1.$$

Probability properties imply that with probability 1 the integral is finite for all n. Then almost certainly for any T, $c > 0$

$$\int_0^T \max_{\|\vec{u}\| \le c} f\left(\vec{u}, \xi(t)\right) dt \le \int_0^n \max_{\|\vec{u}\| \le n} f\left(\vec{u}, \xi(t)\right) dt < \infty,$$

where $n \in \mathcal{N}$; $n \geq T$, $n \geq c$. By Lebesgue theorem of limit transition with probability 1 for any $T > 0$ the function $F_T(\vec{u})$, $\vec{u} \in I$ is continuous. If I is unbounded 7 then by condition 2) and Lemma 4 almost certainly for each $T > 0$

$$F_T(\vec{u}) \to \infty, \quad \| \vec{u} \| \to \infty,$$

and there exists such $\Delta = \Delta(T, \omega) > 0$ that for all $\vec{u} \in I$, $\| \vec{u} \| > \Delta$

$$F_T(\vec{u}) > F_T(\vec{u}_0).$$

Hence with probability 1 for all T

$$\inf_{\vec{u} \in I} F_T(\vec{u}) = \inf_{\vec{u} \in I : \| \vec{u} \| \leq \Delta} F_T(\vec{u}).$$

Then for all $T > 0$ and $\omega \in \Omega'$, $P(\Omega') = 1$ there exists at least one minimum point $\vec{u}(T) = \vec{u}(T, \omega)$ of the function $F_T(\vec{u})$, $\vec{u} \in I$. For any T, $\vec{u}$ the mapping $F_T(\vec{u}, \omega)$, $\omega \in \Omega'$ is $\mathcal{G}_T'$-measurable. Then by Theorem 6 for each $T > 0$ the mapping $\vec{u}(T, \omega)$, $\omega \in \Omega'$ can be chosen $\mathcal{G}_T'$-measurable.

Let us prove that if I is unbounded then there exists such $c > 0$ that, beginning from some T, which depends on ω, all minimum points of the function $F_T(\vec{u})$, $\vec{u} \in I$ belong to the set

$$K_c = \left\{ \vec{u} \in I : \| \vec{u} \| \leq c \right\}$$

with probability 1. By condition 2) with probability 1 for each $t \geq 0$

$$\varphi\left(c, \xi(t)\right) \to \infty, \quad c \to \infty,$$

where

$$\varphi(c, z) = \inf_{\| \vec{u} \| > c} f(\vec{u}, z), \quad z \in Y, \quad c > 0.$$

Hence by Lemma 4

$$E\left\{ \varphi\left(c, \xi(0)\right) \right\} \to \infty, \quad c \to \infty.$$

Let us choose c_0 such that

$$E\left\{ \varphi\left(c_0, \xi(0)\right) \right\} > E\left\{ f\left(\vec{u}_0, \xi(0)\right) \right\}.$$

By properties of ergodic random processes with probability 1, beginning from some T, which depends on ω,

$$\frac{1}{T}\int_0^T \varphi\left(c_0, \xi(t)\right) dt > \frac{1}{T}\int_0^T f\left(\vec{u}_0, \xi(t)\right) dt = F_T(\vec{u}_0)$$

with probability 1. Since

$$\inf_{\|\vec{u}\|>c} F_T(\vec{u}) \geq \frac{1}{T}\int_0^T \varphi\left(c_0, \xi(t)\right) dt,$$

we have proved what we need.

Now we will check the conditions of Theorem 7 for the family of functions

$$\left\{ F_T : K \times \Omega' \to \mathcal{R}, \quad T > 0 \right\}, \tag{2.20}$$

where $K = K_{c_0}$.

It is evident that conditions 1) and 2) of Theorem 7 are satisfied. Since the process $\xi(t)$ is ergodic, then

$$P\left\{ \lim_{T\to\infty} F_T(\vec{u}) = F(\vec{u}) \right\} = 1, \quad \vec{u} \in I.$$

By condition 1) the function $F(\vec{u})$ is continuous. Then condition 3) of Theorem 7 is fulfilled.

Denote

$$\psi(\gamma, z) = \sup_{\vec{u}, \vec{v} \in K : \|\vec{u} - \vec{v}\| < \gamma} \left| f(\vec{u}, z) - f(\vec{v}, z) \right|, \quad \gamma > 0, \quad z \in Y.$$

With probability 1 for all $T > 0$, $\vec{u}_1 \in K$, $\gamma > 0$

$$\zeta_T(\vec{u}_1, \gamma) = \sup_{\vec{u} \in K : \|\vec{u} - \vec{u}_1\| < \gamma} \left| F_T(\vec{u}) - F_T(\vec{u}_1) \right| \leq \frac{1}{T}\int_0^T \psi\left(\gamma, \xi(t)\right) dt. \tag{2.21}$$

By properties of ergodic processes

$$P\left\{ \lim_{T\to\infty} \frac{1}{T}\int_0^T \psi\left(\gamma, \xi(t)\right) dt = E\left\{ \psi\left(\gamma, \xi(0)\right) \right\} \right\} = 1, \quad \gamma > 0. \tag{2.22}$$

Denote

$$c(\gamma) = E\left\{\psi\big(\gamma, \xi(0)\big)\right\} + \gamma, \quad \gamma > 0.$$

By (2.21), (2.22) for any $\vec{u}_1 \in K$, $\gamma > 0$

$$P\left\{\overline{\lim_{T \to \infty}} \zeta_T(\vec{u}_1, \gamma) < c(\gamma)\right\} = 1.$$

As in Theorem 8, 10

$$c(\gamma) \to 0, \quad \gamma \to 0.$$

Then the proof is complete.

As for a process with discrete time we will make a note. Let $\left\{\vec{\eta}(t), t \in \mathcal{R}\right\}$ be a measurable stationary in a strict sense random process, defined on a complete probabilistic space $(\Omega, \mathcal{G}, P)$, with values in $\mathcal{R}^m$, $m \geq 1$. Suppose that the process $\vec{\eta}(t)$ is stochastically continuous and satisfies a strong mixing condition with a coefficient $\alpha(\tau) = O(\tau^{-1-\varepsilon})$, $\tau \to \infty$; $\varepsilon > 0$; and $E\left\{\|\vec{\eta}(0)\|^{2+\delta}\right\} < \infty$, where $\delta > 4/\varepsilon$. Then the process $\vec{\eta}(t)$ has a spectral density which is bounded and continuous function [54].

Theorem 13 *Let traectories of the process $\xi(t)$ be continuous with probability 1 and the function f be continuous on $I \times Y$ and*

$$E\left\{f\big(\vec{u}, \xi(0)\big)\right\} < \infty, \quad \vec{u} \in I.$$

Suppose that $\vec{u}_0$ and $\vec{u}(T) = \vec{u}(T, \omega)$ are some minimum points of the functions (2.18) and (2.19) respectively; $T > 0$, $\omega \in \Omega'$, $P(\Omega') = 1$; for any $T > 0$ the mapping $\vec{u}(T, \omega)$, $\omega \in \Omega'$ is $\mathcal{G}'_T$-measurable; $\vec{u}(T) \to \vec{u}_0$, $T \to \infty$ with probability 1; and the following conditions are satisfied:

1) *$\vec{u}_0$ is an internal point of I;*

2) *there exists such a closed neighbourhood S of $\vec{u}_0$ that for each $z \in Y$ the function $f(\vec{u}, z)$, $\vec{u} \in S$ is twice continuously differentiable on S;*

3) *$E\left\{\max_{\vec{u} \in S}\left\|\vec{\nabla} f\big(\vec{u}, \xi(0)\big)\right\|\right\} < \infty$, where*

$$\vec{\nabla} f(\vec{u}, z) = \left(\frac{\partial f}{\partial u_j}(\vec{u}, z)\right)^l_{j=1}, \quad \vec{u} = (u_j)^l_{j=1} \in S, \quad z \in Y;$$

4) for any $T > 0$ with probability 1

$$\int\limits_0^T \max_{\vec{u} \in S} \| \Phi \left(\vec{u}, \xi(t) \right) \| dt < \infty,$$

where

$$\Phi(\vec{u}, z) = \left(\frac{\partial^2 f}{\partial u_j \partial u_k} (\vec{u}, z) \right)_{j,k=1}^l \; ;$$

5) $E \left\{ \max_{\vec{u} \in S} \| \Phi \left(\vec{u}, \xi(0) \right) \| \right\} < \infty;$

6) $\det A_0 \neq 0$, where $A_0 = E \left\{ \Phi \left(\vec{u}_0, \xi(0) \right) \right\};$

7) the process $\xi(t)$ satisfies a strong mixing condition with the coefficient $\alpha(\tau) = O(\tau^{-1-\varepsilon})$, $\tau \to \infty$, where $\varepsilon > 0$;

8) for some $\delta > 4/\varepsilon$

$$E \left\{ \left\| \vec{\nabla} f \left(\vec{u}_0, \xi(0) \right) \right\|^{2+\delta} \right\} < \infty;$$

9) the function $\vec{\nabla} f(\vec{u}_0, z)$, $z \in Y$ is continuous;

10) $\det g(0) \neq 0$, where $g(\lambda) = (g_{jk}(\lambda))_{j,k=1}^l$ *is the spectral density of the process* $\left\{ \vec{\nabla} f \left(\vec{u}_0, \xi(t) \right), t \in \mathcal{R} \right\}.$

Then $\sqrt{T} \left(\vec{u} \, (T) - \vec{u}_0 \right)$ *converges in distribution to* $\mathcal{N} \left(\vec{0}, 2\pi (A_0)^{-1} g(0) (A_0)^{-1} \right)$, $T \to \infty$.

Assume that the following conditions also take place:

11) for some $\delta_1 > 4/\varepsilon$

$$E \left\{ \left(f \left(\vec{u}_0, \xi(0) \right) \right)^{2+\delta_1} \right\} < \infty;$$

12) $g_1(0) \neq 0$, where $g_1(\lambda)$ *is the spectral density of the process* $\left\{ f \left(\vec{u}_0, \xi(t) \right), t \in \mathcal{R} \right\}.$

Then $\sqrt{T}\left(F_T\left(\vec{u}\left(T\right)\right) - F(\vec{u}_0)\right)$ converges in distribution to $\mathcal{N}\left(0, 2\pi g_1(0)\right)$, $T \to \infty$.

Proof. By condition 3) and Fubini theorem with probability 1 for any $T > 0$

$$\int\limits_0^T \max_{\vec{u} \in S} \| \vec{\nabla} f\left(\vec{u}, \xi(t)\right) \| dt < \infty.$$

Then condition 2), Tailor formula and Lebesgue theorem of limit transition imply that with probability 1 for each $T > 0$ the function $F_T(\vec{u})$, $\vec{u} \in S$ is differentiable on S and

$$\vec{\nabla} F_T(\vec{u}) = \left(\frac{\partial F_T}{\partial u_j}(\vec{u})\right)_{j=1}^l = \frac{1}{T}\int\limits_0^T \vec{\nabla} f\left(\vec{u}, \xi(t)\right) dt, \quad \vec{u} \in S.$$

By conditions 2) and 4) the function $F_T(\vec{u})$, $\vec{u} \in S$ is twice continuously differentiable and

$$\frac{\partial^2 F_T}{\partial u_j \partial u_k}(\vec{u}) = \frac{1}{T}\int\limits_0^T \frac{\partial^2 f}{\partial u_j \partial u_k}\left(\vec{u}, \xi(t)\right) dt, \quad j, k = \overline{1, l}, \quad \vec{u} \in S$$

almost surely for all $T > 0$.

As in Theorem 9 and Theorem 11 for any $T > 0$, $\omega \in \Omega(T)$, $P\left(\Omega(T)\right) \to 1$, $T \to \infty$ we have

$$\vec{\nabla} F_T(\vec{u}_0) + A(T)\left(\vec{u}\left(T\right) - \vec{u}_0\right) = \vec{0}, \tag{2.23}$$

where

$$A(T) = A(T, \omega) = \left(\frac{\partial^2 F_T}{\partial u_j \partial u_k}\left(\vec{u}^j\left(T, \omega\right), \omega\right)\right)_{j,k=1}^l;$$

$$\vec{u}^j\left(T, \omega\right) = \vec{u} + \theta_j\left(T, \omega\right)\left(\vec{u}^j\left(T, \omega\right) - \vec{u}_0\right); \quad 0 < \theta_j\left(T, \omega\right) < 1,$$

and for each $T > 0$ the mapping $\theta_j\left(T, \omega\right)$, $\omega \in \Omega(T)$ is $\mathcal{G}$-measurable.

Denote

$$A_1(T) = A_1(T,\omega) = \begin{cases} A(T,\omega), & \omega \in \Omega(T) \\ \\ A_0, & \omega \in \Omega \setminus \Omega(T) \end{cases}, \quad T > 0, \quad \omega \in \Omega.$$

As in Theorem 9 for any $T > 0$, $\omega \in \Omega_1(T)$, $P(\Omega_1(T)) \to 1$, $T \to \infty$ the equation (2.23) can be rewritten in such a way:

$$\sqrt{T}\left(\vec{u}(T) - \vec{u}_0\right) = -(A_1(T))^{-1} \frac{1}{\sqrt{T}} \int_0^T \vec{\nabla} f\left(\vec{u}_0, \xi(t)\right) dt. \qquad (2.24)$$

By conditions 7) – 10) and Theorem 4 we have that

$$\frac{1}{\sqrt{T}} \int_0^T \vec{\nabla} f\left(\vec{u}_0, \xi(t)\right) dt$$

converges in distribution to $\mathcal{N}\left(\vec{0}, 2\pi\, g(0)\right)$, $n \to \infty$.

Now as in Theorem 11 we have due to (2.24) the proof of the first part of Theorem 13.

As in Theorem 9 and Theorem 11 for any $T > 0$, $\omega \in \Omega(T)$, $P(\Omega(T)) \to 1$, $T \to \infty$ we have

$$\sqrt{T}\left(F_T\left(\vec{u}(T)\right) - F(\vec{u}_0)\right) = \left(\frac{1}{\sqrt{T}} \int_0^T \vec{\nabla} f\left(\vec{u}_0, \xi(t)\right) dt\right)' \times$$

$$\times \left(\vec{u}(T) - \vec{u}_0\right) + \frac{\sqrt{T}}{2}\left(\vec{u}(T) - \vec{u}_0\right)' A_3(T)\left(\vec{u}(T) - \vec{u}_0\right) +$$

$$+ \frac{1}{\sqrt{T}} \int_0^T \left(f\left(\vec{u}_0, \xi(t)\right) - E\left\{f\left(\vec{u}_0, \xi(0)\right)\right\}\right) dt, \qquad (2.25)$$

where

$$A_3(T) = A_3(T,\omega) = \begin{cases} \left(\frac{\partial^2 F_T}{\partial u_j \partial u_k}(\vec{v}(T,\omega),\omega)\right)_{j,k=1}^l, & \omega \in \Omega(T) \\ \\ A_0, & \omega \in \Omega \setminus \Omega(T) \end{cases},$$

$$T > 0, \quad \omega \in \Omega,$$

$$\vec{v}\,(T) = \vec{v}\,(T, \omega) = \vec{u}_0 + \theta(T, \omega)\left(\vec{u}\,(T, \omega) - \vec{u}_0\right); \quad 0 < \theta(T, \omega) < 1,$$

and for each $T > 0$ the mapping $\theta(T, \omega)$, $\omega \in \Omega(T)$ is $\mathcal{G}$-measurable.

By conditions 7), 11), 12) and Theorem 4

$$\frac{1}{\sqrt{T}} \int\limits_0^T \left(f\left(\vec{u}_0, \xi(t)\right) - E\left\{ f\left(\vec{u}_0, \xi(0)\right)\right\}\right) dt$$

converges in distribution to $\mathcal{N}\left(\vec{0}, 2\pi g_1(0)\right)$, $T \to \infty$. Then as in Theorems 9 and 11 the equation (2.25) implies the validity of the second part of Theorem 13.

2.4 MODELS WITH RESTRICTIONS IN THE FORM OF INEQUALITIES

When look for an asymptotic distribution of empirical estimates for stochastic programming problems, considered above, we assumed a minimum point of the criterion function to be an internal point of a set of available solutions. Consider now a case when this condition is not satisfied.

We will use the methods formulated in [90].

Assume that $\{\xi_i, i \geq 1\}$ is a sequence of independent identically distributed random variables defined on the probabilistic space $(\Omega, \mathcal{G}, P)$, with values in $\mathcal{R}$. The set I is defined as

$$I = \left\{ \vec{x} \in \mathcal{R}^l : \vec{g}\,(\vec{x}) = \left(g_1(\vec{x}), \ldots, g_m(\vec{x})\right) \leq \vec{0} \right\},$$

where $m \geq 1$, $l \geq 1$, " $\leq$ " applies to each component. Let the function $f : I \times \mathcal{R} \to \mathcal{R}_+$ be continuous in the first argument and measurable in the second argument. Define the set of functionals

$$F_n(\vec{x}) = \frac{1}{n} \sum_{i=1}^n f(\vec{x}, \xi_i), \quad \vec{x} \in I. \tag{2.26}$$

We assume that the following conditions are satisfied.

1. There exists the point $\vec{x}^*$ such that

$$Ef(\vec{x},\xi_1) > Ef(\vec{x}^*,\xi_1), \quad \vec{x} \in I, \quad \vec{x} \neq \vec{x}^*$$

and

$$P\left\{ \lim_{n \to \infty} \|\vec{x}_n - \vec{x}^*\| = 0 \right\} = 1, \quad \vec{x}_n \in \underset{\vec{x} \in I}{\text{Arg min}} \, F_n(\vec{x}).$$

2. The function $f(\vec{x},y)$, $\vec{x} \in I$ is twice continuously differentiable for each fixed $y \in \mathcal{R}$, and

$$E \max_{\|\vec{x}\| \leq c} f(\vec{x},\xi_1) < \infty, \quad c > 0.$$

There exist $c > 0$, $\gamma_0 > 0$ such that if $\|\vec{x} - \vec{x}^*\| \leq \gamma$, $\gamma < \gamma_0$, then we have the inequalities

$$E\left| \frac{\partial f}{\partial x_j}(\vec{x},\xi_1) \right| \leq c, \quad E\left| \frac{\partial^2 f(\vec{x},\xi_1)}{\partial x_j \partial x_k} \right| \leq c, \quad j,k = \overline{1,l}, \quad \vec{x} = (x_k)_{k=1}^l.$$

The matrix

$$\Phi(\vec{x}^*) = \left(E \frac{\partial^2 f(\vec{x}^*,\xi_1)}{\partial x_j \partial x_k} \right)_{j,k=1}^l$$

is positive definite.

3. The functions $g_i(\vec{x})$, $\vec{x} \in I$ are twice continuously differentiable, and for some $c > 0$ $\|\vec{x} - \vec{x}^*\| \leq \gamma$, $\gamma < \gamma_0$

$$\left| \frac{\partial g_i(\vec{x})}{\partial x_k} \right| \leq c, \quad \left| \frac{\partial^2 g_i(\vec{x})}{\partial x_j \partial x_k} \right| \leq c, \quad j,k = \overline{1,l}.$$

4. N_1 is the set of indexes i, for which $g_i(\vec{x}^*) = 0$, and N_2 is the set of indexes i, for which $g_i(\vec{x}^*) < 0$. We assume that

$$\vec{\nabla} g_i(\vec{x}^*) = \left(\frac{\partial g_i}{\partial x_k}(\vec{x}^*) \right)_{k=1}^l, \quad i \in N_1$$

are linearly independent.

5. The functions $g_i(\vec{x})$, $\vec{x} \in I$, $i = \overline{1,m}$ are convex.

6. There exists a point $\vec{x}_*$ such that $\vec{g}(\vec{x}_*) < \vec{0}$.

As noted in [90], conditions 5 and 6 imply that $\vec{\nabla} g_i(\vec{x}_n)$ are linearly independent for those i where $g_i(\vec{x}_n) = 0$.

Let us now investigate the asymptotic distribution of the variables $\vec{x}_n$. We introduce some relationships that follow from the necessary conditions of extremum. By definition

$$\vec{\nabla} F_n(\vec{x}_n) + \sum_{i=1}^{m} \lambda_{in} \vec{\nabla} g_i(\vec{x}_n) = \vec{0},$$

$$\lambda_{in} g_i(\vec{x}_n) = 0, \quad \lambda_{in} \geq 0, \quad i = \overline{1,m}, \tag{2.27}$$

where λ_{in} are Lagrange multipliers. Denote $\vec{\nabla} F_n(\vec{x}) = \widetilde{F}_n(\vec{x})$. Applying the mean value theorem, we obtain

$$\widetilde{F}_n(\vec{x}_n) = \widetilde{F}_n(\vec{x}^*) + \widetilde{\Phi}_n(\vec{x}_n - \vec{x}^*), \tag{2.28}$$

where $\widetilde{\Phi}_n$ is a $l \times l$ matrix with elements $\widetilde{\Phi}_n^{jk}$ of the form

$$\widetilde{\Phi}_n^{jk} = \frac{1}{n} \sum_{i=1}^{n} \frac{\partial^2 f(\vec{x}, \xi_i)}{\partial x_j \partial x_k} \bigg|_{\vec{x} = \vec{\zeta}_{jn}},$$

$$\vec{\zeta}_{jn} = \vec{x}^* + \theta_j(\vec{x}_n - \vec{x}^*), \quad 0 \leq \theta_j \leq 1, \quad j = \overline{1,l}.$$

We thus have

$$\widetilde{F}_n(\vec{x}^*) + \widetilde{\Phi}_n(\vec{x}_n - \vec{x}^*) + \sum_{i=1}^{m} \lambda_{in} \vec{\nabla} g_i(\vec{x}_n) = \vec{0}. \tag{2.29}$$

Let us prove some lemmas.

Lemma 7 *Assume that condition 2 holds. Then*

$$P\left\{ \lim_{n \to \infty} \Phi_n^{jk}(\vec{x}_n) = \Phi^{jk}(\vec{x}^*) \right\} = 1, \quad j, k = \overline{1,l},$$

where $\| \vec{x}_n - \vec{x}^* \| \to 0$, $n \to \infty$ *with probability* 1,

$$\Phi_n^{jk}(\vec{x}) = \frac{\partial^2 F_n(\vec{x})}{\partial x_j \partial x_k}, \quad \Phi^{jk}(\vec{x}) = E \frac{\partial^2 f(\vec{x}, \xi_1)}{\partial x_j \partial x_k}.$$

Proof. Take

$$\sup_{\vec{x} \,:\, \|\vec{x} - \vec{x}^*\| < \gamma} \left| \Phi_n^{jk}(\vec{x}) - \Phi_n^{jk}(\vec{x}^*) \right| \le$$

$$\le \frac{1}{n} \sum_{i=1}^{n} \sup_{\vec{x} \,:\, \|\vec{x} - \vec{x}^*\| < \gamma} \left| \frac{\partial^2 f(\vec{x}, \xi_i)}{\partial x_j \partial x_k} - \frac{\partial^2 f(\vec{x}^*, \xi_i)}{\partial x_j \partial x_k} \right| = \zeta_n^{jk}(\gamma).$$

Let

$$c_{jk}(\gamma) = 2E \left\{ \sup_{\vec{x} \,:\, \|\vec{x} - \vec{x}^*\| < \gamma} \left| \Phi^{jk}(\vec{x}) - \Phi^{jk}(\vec{x}^*) \right| \right\}.$$

By Theorem 1

$$P \left\{ \lim_{n \to \infty} \zeta_n^{jk}(\gamma) = \frac{1}{2} c_{jk}(\gamma) \right\} = 1.$$

Since the function $\partial^2 f(\vec{x}^*, \xi_1)/\partial x_j \partial x_k$ is continuous in $\vec{x}$, we pass to the limit under the expectation symbol and obtain that $c(\gamma) \to 0$ as $\gamma \to 0$. Because of Theorem 1 for each $\vec{x}$

$$P \left\{ \lim_{n \to \infty} \Phi_n^{jk}(\vec{x}) = \Phi^{jk}(\vec{x}) \right\} = 1.$$

Thus the conditions of Lemma 5 are satisfied, and the proof is complete.

Let us note that by Lemma 7

$$P \left\{ \lim_{n \to \infty} \tilde{\Phi}_n^{jk} = \Phi^{jk}(\vec{x}^*) \right\} = 1.$$

Now we shall find the asymptotic distribution of the optimum points. Multiply both sides (2.29) by $\sqrt{n}$ and apply the mean value theorem to the functions $g_i(\vec{x})$:

$$\sqrt{n}\, \tilde{F}_n(\vec{x}^*) + \tilde{\Phi}_n \sqrt{n}\, (\vec{x}_n - \vec{x}^*) + \sum_{i=1}^{m} \sqrt{n}\, \lambda_{in}\, \vec{\nabla}\, g_i(\vec{x}_n) = \vec{0}, \qquad (2.30)$$

$$\lambda_{in}\, g_i(\vec{x}^{\,*}) + \lambda_{in}\, \vec{\nabla}\, g_i\left(\vec{x}^{\,*} + \theta_i(\vec{x}_n - \vec{x}^{\,*})\right)(\vec{x}_n - \vec{x}^{\,*}) = 0, \quad i = \overline{1,m}. \quad (2.31)$$

Thus we have the following quadratic programming problem:

$$\varphi_n(\vec{x}) = \frac{1}{2}\,\vec{x}^{\,\prime}\,\tilde{\Phi}_n\,\vec{x} + \tilde{F}'_n(\vec{x}^{\,*})\,\vec{x} \to \min, \quad (2.32)$$

$$b_i(\vec{x}) = \sqrt{n}\, g_i(\vec{x}^{\,*}) + \vec{\nabla}\, g_i(\vec{x}^{\,*})\,\vec{x} + l_i(\vec{x}) \le 0,$$

$$\vec{x} \in \mathcal{R}^l, \quad l_i(\vec{x}) = \frac{1}{2\sqrt{n}}\,\vec{x}^{\,\prime}\,\tilde{g}_i(\vec{x})\,\vec{x}, \quad i = \overline{1,m}, \quad (2.33)$$

where $\tilde{g}_i(\vec{x}) = \left(\dfrac{\partial^2 g_i}{\partial x_j \partial x_k}\left(\vec{x}^{\,*} + \theta_{ij}\dfrac{\vec{x}}{\sqrt{n}}\right)\right)^l_{j,k=1}.$

By the central limit theorem $\sqrt{n}\,\tilde{F}_n(\vec{x}^{\,*})$ converges weakly to the $\mathcal{N}(\vec{0}, B)$ – distributed variable, where

$$B = E\left\{\vec{\nabla}\, f(\vec{x}^{\,*}, \xi_1)\left(\vec{\nabla}\, f(\vec{x}^{\,*}, \xi_1)\right)'\right\}.$$

Alongside (2.32) - (2.33), consider the following quadratic programming problem:

$$\frac{1}{2}\,\vec{x}^{\,\prime}\,\Phi(\vec{x}^{\,*})\,\vec{x} + \vec{\xi}^{\,\prime}\,\vec{x} \to \min, \quad (2.34)$$

$$\vec{\nabla}\, g'_i(\vec{x}^{\,*})\,\vec{x} \le 0, \quad i \in N_1, \quad (2.35)$$

where $\vec{\xi}$ is the $\mathcal{N}(\vec{0}, B)$ – distributed vector.

Theorem 14 *Assume that conditions $1-6$ are satisfied. Then the random vector $\vec{\zeta}_n = \sqrt{n}(\vec{x}_n - \vec{x}^{\,*})$, solving the problem (2.32) - (2.33), tends in distribution to the random vector $\vec{\zeta}$ that solves the problem (2.34) - (2.35).*

Corollary 1 *If $N_1 = \oslash$ then the vector $\vec{\zeta}_n$ converges weakly to*

$$\mathcal{N}\left(\vec{0}, \Phi^{-1}(\vec{x}^{\,*})\, B\, \Phi^{-1}(\vec{x}^{\,*})\right), \quad n \to \infty.$$

The proof of Theorem 14 essentially exploites the following lemmas.

Denote

$$\vec{v}_n = \sqrt{n}\,\vec{\lambda}_n, \quad \vec{v}_n = (\vec{v}_{1n}, \vec{v}_{2n}) \in \mathcal{R}^m, \quad \vec{\lambda}_n = (\lambda_{in})_{i=1}^m,$$

where $\vec{v}_{1n}$ – a part of $\vec{v}_n$ with indexes from N_1, $\vec{v}_{2n}$ – with indexes from N_2.

Lemma 8 [90] *A quadratic programming problem*

$$\min\left\{\frac{1}{2}\,\vec{x}'\,R\,\vec{x} - \vec{q}'\,\vec{x} \,\middle|\, \vec{g}_i'\vec{x} \leq 0, \quad i \in I\right\}$$

has the solution $\vec{x}\,(\vec{q})$, continuous in $\vec{q}$. Here $\vec{x}$, $\vec{q}$, $\vec{g}_i \in \mathcal{R}^n$, R – a positive definite matrix $n \times n$.

Lemma 9 [90]. *Let A_n be the random symmetric matrix $m \times m$, converging in probability to the positive definite matrix A, $n \to \infty$. Then*

1. *a_{kn}, $k = \overline{1,m}$ – eigenvalues of A_n converge in probability to a_k, $k = \overline{1,m}$ – eigenvalues of A;*

2. *$\lim\limits_{n\to\infty} p_n = 1$, where p_n is probability of an event, that A_n has positive eigenvalues.*

Lemma 10 *Assume that conditions $1, 2$ are satisfied. Then*

$$P\left\{\lim_{n\to\infty} \vec{v}_{2n} = \vec{0}\right\} = 1.$$

Proof. Because of the continuity of functions g_i

$$P\left\{\lim_{n\to\infty} g_i(\vec{x}_n) = g_i(\vec{x}^*)\right\} = 1, \quad i = \overline{1,m}.$$

Now the lemma follows from (2.25) due to

$$g_i(\vec{x}^*) < 0, \quad i \in N_2.$$

Lemma 11 *Suppose that conditions* $1-6$ *are fulfilled. Then for a given* $\delta > 0$ *there exist* $\varepsilon > 0$ *and* $N_0 > 0$ *such that*

$$P\left\{\| \vec{v}_{1n} \| > \varepsilon \right\} < \delta, \quad n > N_0. \tag{2.36}$$

Proof. Denote $\widetilde{\widetilde{\Phi}}_n = \left(\widetilde{\Phi}_n + \gamma_n(| \wedge_{1n} | + c)I \right)^{-1}$ the matrix, defined in such a way: $\gamma_n = 1$, if $\det \widetilde{\Phi}_n = 0$, and else $\gamma_n = 0$; $\wedge_{1n}$ is a minimal eigenvalue of $\widetilde{\Phi}_n$, $c > 0$, $I = (i_{jk})_{j,k=1}^l$, $i_{jk} = 1$, $j = k$; $i_{jk} = 0$, $j \neq k$.

It is easy to prove that

$$P\left\{ \widetilde{\widetilde{\Phi}}_n \widetilde{\Phi}_n = I \right\} \geq P\left\{ \gamma_n = 0 \right\} = P\left\{ \det \widetilde{\Phi}_n \neq 0 \right\} \geq 1 - \delta_0, \quad n > n_0,$$

where δ_0 is an arbitrary positive number. Then from (2.30) we have

$$P\left\{ \widetilde{\widetilde{\Phi}}_n \widetilde{\Phi}_n = I \right\} = P\left\{ \vec{\zeta}_n = -\widetilde{\widetilde{\Phi}}_n \sqrt{n}\, \vec{F}_n(\vec{x}^*) - \widetilde{\widetilde{\Phi}}_n G_1'(\vec{x}_n)\, \vec{v}_{1n} - \right.$$

$$\left. -\widetilde{\widetilde{\Phi}}_n G_2'(\vec{x}_n)\, \vec{v}_{2n} \right\} \geq 1 - \delta_0, \quad n > n_0,$$

where $G_1(\vec{x})$ is the $m_1 \times l$-matrix with i-th row $\vec{\nabla} \vec{g}_i'(\vec{x})$, $i \in N_1$, $G_2(\vec{x})$ – the $(m - m_1) \times l$-matrix, $\vec{\nabla} \vec{g}_i'(\vec{x})$, $i \in N_2$ – its i-th row, m_1 – the number of indexes in N_1.

Denote

$$B_n = \psi_1(\vec{x}_n)\, \widetilde{\widetilde{\Phi}}_n G_1'(\vec{x}_n), \quad \vec{q}_n = -\psi_1(\vec{x}_n)\, \widetilde{\widetilde{\Phi}}_n \sqrt{n}\, \vec{F}_n(\vec{x}^*),$$

$$\vec{h}_n = \psi_1(\vec{x}_n)\, \widetilde{\widetilde{\Phi}}_n G_2'(\vec{x}_n)\, \vec{v}_{2n},$$

where $\psi_1(\vec{x})$ is the $m_1 \times l$-matrix with i-th row $\psi_i(\vec{x})$, $i \in N_1$;

$$\psi_i(\vec{x}) = \vec{\nabla} g_i'(\vec{x}^*) + \frac{1}{2}(\vec{x} - \vec{x}^*)' \widetilde{g}_i(\vec{x} - \vec{x}^*).$$

We have

$$\vec{v}\,'_{1n}\, B_n\, \vec{v}_{1n} = \vec{v}\,'_{1n}\, A_n\, \vec{v}_{1n},$$

where A_n is the symmetric matrix with elements

$$a_{ij}^n = \frac{1}{2}\left(b_{ij}^n + b_{ji}^n\right), \quad i,j = \overline{1,l},$$

where b_{ij}^n – elements of B_n. Hence,

$$P\left\{\tilde{\tilde{\Phi}}_n \tilde{\Phi}_n = I\right\} = P\left\{\vec{v}\,'_{1n}\, A_n\, \vec{v}_{1n} + \vec{v}\,'_{1n}\, (\vec{h}_n - \vec{q}_n) = 0\right\} \geq 1 - \delta_0, \quad n > n_0.$$

For the symmetric matrix A_n we have

$$C_n'\, A_n\, C_n = M_n,$$

where C_n is an orthogonal matrix $(C_n' C_n = I)$, $M_n = \text{diag}\,(\nu_{1n}, \ldots, \nu_{ln})$, ν_{in} – i-th eigenvalue of A_n. By Lemma 7

$$p \lim_{n \to \infty} \tilde{\tilde{\Phi}}_n = \Phi^{-1}(\vec{x}^*).$$

Because of the consistency of $\vec{x}_n$ and condition 3

$$p \lim_{n \to \infty} \psi_1(\vec{x}_n) = p \lim_{n \to \infty} G_1(\vec{x}_n) = G_1(\vec{x}^*).$$

Then

$$p \lim_{n \to \infty} B_n = p \lim_{n \to \infty} A_n = G_1(\vec{x}^*) \Phi^{-1}(\vec{x}^*) G_1(\vec{x}^*) = A.$$

Then, conditions $2, 4$ imply that A is positively defined. From Lemma 9

$$p \lim_{n \to \infty} M_n = M, \quad p \lim_{n \to \infty} C_n = C,$$

where $M = \text{diag}\,(\nu_1, \ldots, \nu_l)$, ν_i – i-th eigenvalue of A, C – an orthogonal matrix such that $C'AC = M$.

Define $\tilde{\nu}_{in} = \nu_{in}$, $\nu_{in} > 0$; else $\tilde{\nu}_{in} = 1$. We have

$$p \lim_{n \to \infty} \widetilde{M}_n = M,$$

where $\widetilde{M}_n = \mathrm{diag}(\widetilde{\nu}_{1n}, \ldots, \widetilde{\nu}_{ln})$. By Lemma 9 for any $\delta' > 0$ there exists such n_1 that

$$P\left\{M_n = \widetilde{M}_n\right\} = P\left\{\nu_{in} > 0, \ i = \overline{1,l}\right\} > 1 - \frac{\delta'}{3}, \quad n > n_1.$$

Denote $\vec{Y}_n = \widetilde{M}_n^{1/2} C_n^{-1} \vec{v}_{1n}$. Because of the inequality

$$P\left\{\det \widetilde{\Phi}_n = 0\right\} \leq \delta_0, \quad n > n_0,$$

we obtain

$$P\left\{\vec{Y}_n'\vec{Y}_n + 2\,\vec{Y}_n'\vec{K}_n = 0\right\} \geq P\left\{\nu_{in} > 0, \quad i = \overline{1,l}; \quad \det \widetilde{\Phi}_n \neq 0\right\} =$$

$$= P\left\{\nu_{in} > 0, \quad i = \overline{1,l}\right\} - P\left\{\nu_{in} > 0, \quad i = \overline{1,l}; \quad \det \widetilde{\Phi}_n = 0\right\} \geq$$

$$\geq P\left\{\nu_{in} > 0, \ i = \overline{1,l}\right\} - P\left\{\det \widetilde{\Phi}_n = 0\right\} \geq 1 - \frac{\delta}{3}, \quad n > n_2 = \max(n_0, n_1),$$

where $\delta = 3\delta_0 + \delta'$, $\vec{K}_n = \frac{1}{2}\widetilde{M}_n^{-1/2} C_n (\vec{h}_n - \vec{q}_n)$.

In such a way we have

$$P\left\{\vec{Y}_n'\vec{Y}_n + 2\,\vec{Y}_n'\vec{K}_n = 0\right\} \geq 1 - \frac{\delta}{3}, \quad n > n_2.$$

As it was shown above

$$-\sqrt{n}\,\widetilde{F}_n(\vec{x}^*) \Longrightarrow \vec{Q}, \quad n \to \infty,$$

where $\vec{Q}$ is the normally distributed random vector. Then we have

$$\vec{q}_n \Rightarrow \vec{q}, \quad n \to \infty; \quad p\lim_{n \to \infty} \vec{h}_n = \vec{0} \ .$$

Thus limit distribution of $\vec{K}_n$ coincides with the distribution of the random vector $\vec{K} = -\frac{1}{2}M^{-1/2}C'\,\vec{q}$. It is easy to show that components of the m_1-dimension random variable $2\,\vec{K}$ are independent $\mathcal{N}(0,1)$ – distributed random variables. This implies that $4\| \vec{K}_n \|^2$ has the limit distribution χ^2. Define $\vec{Y}_n = \widetilde{Y}_n - \vec{K}_n$. Then

$$P\left\{\|\widetilde{Y}_n\| = \| \vec{K}_n \|\right\} \geq 1 - \frac{\delta}{3}, \quad n > n_2.$$

The random vector $\vec{K}_n$ has the limit distribution, then for any $\delta > 0$ we can find n_3 and $\varepsilon_1 > 0$ such that

$$\frac{\delta}{6} > P\left\{\| \vec{K}_n \| \geq \frac{\varepsilon_1}{2}\right\} \geq P\left\{\|\widetilde{Y}_n\| \geq \frac{\varepsilon_1}{2}, \quad \| \vec{K}_n \| \geq \frac{\varepsilon_1}{2}\right\} \geq$$

$$\geq P\left\{\|\widetilde{Y}_n\| \geq \frac{\varepsilon_1}{2}\right\} - P\left\{\|\widetilde{Y}_n\| \neq \| \vec{K}_n \|\right\}, \quad n > n_3.$$

Hence

$$P\left\{\|\widetilde{Y}_n\| \geq \frac{\varepsilon_1}{2}\right\} < \frac{\delta}{2}, \quad n > n_4 = \max(n_2, n_3).$$

It is evident that

$$\| \vec{Y}_n \| \leq \|\widetilde{Y}_n\| + \| \vec{K}_n \|.$$

Then for any $\delta > 0$ there exists such $\varepsilon_1 > 0$ that

$$P\left\{\| \vec{Y}_n \| < \varepsilon_1\right\} \geq P\left\{\|\widetilde{Y}_n\| < \frac{\varepsilon_1}{2}\right\} - P\left\{\| \vec{K}_n \| \geq \frac{\varepsilon_1}{2}\right\} > 1 - \frac{2\delta}{3}, \quad n > n_4.$$

Then

$$\| \vec{v}_{1n} \| \leq \|C_{1n}\| \| \vec{Y}_n \|,$$

where $C_{1n} = C_n M_n^{-1/2}$. We have

$$p \lim_{n \to \infty} C_{1n} = C_1 = CM^{-1/2}.$$

This implies that for each $\delta > 0$ there exists such $\varepsilon > 0$ that

$$P\left\{\|C_{1n}\| \varepsilon_1 < \varepsilon\right\} \geq 1 - \frac{\delta}{3}, \quad n > n_5.$$

Denote $n_6 = \max(n_4, n_5)$. If for the inequality

$$P\left\{a \geq b\right\} \geq P\left\{a \geq \varsigma\right\} - P\left\{b \geq \varsigma\right\} \tag{2.37}$$

we let $a = \|C_{1n}\| \| \vec{Y}_n \|$, $b = \|C_{1n}\| \varepsilon_1$, $\zeta = \varepsilon$, we obtain

$$\frac{2}{3}\delta > P\left\{\|C_{1n}\| \| \vec{Y}_n \| \geq \|C_{1n}\| \varepsilon_1\right\} \geq P\left\{\|C_{1n}\| \| \vec{Y}_n \| \geq \varepsilon\right\} - \frac{\delta}{3}, \quad n > n_6.$$

This implies the validity of the lemma.

Lemma 12 *Let conditions* $1 - 6$ *be fulfilled. Then for given* $\delta > 0$ *there exist* $\varepsilon > 0$ *and* N_0 *such that*

$$P\left\{\|\sqrt{n}\,(\vec{x}_n - \vec{x}^*)\| > \varepsilon\right\} < \delta, \quad n > N_0.$$

Proof. Define

$$\vec{\Sigma}_n = -\widetilde{\widetilde{\Phi}}_n \sqrt{n}\,\widetilde{F}_n(\vec{x}^*) - \widetilde{\widetilde{\Phi}}_n\, G_1'(\vec{x}_n)\,\vec{v}_{1n} - \widetilde{\widetilde{\Phi}}_n\, G_2'(\vec{x}_n)\,\vec{v}_{2n}\,.$$

For some $\varepsilon_3 > 0$ we have

$$P\left\{\| \vec{\zeta}_n \| \geq \varepsilon_3, \| \vec{\zeta}_n \| = \| \vec{\Sigma}_n \|\right\} \geq P\left\{\| \vec{\zeta}_n \| \geq \varepsilon_3\right\} - P\left\{\| \vec{\zeta}_n \| \neq \| \vec{\Sigma}_n \|\right\}.$$

Hence

$$P\left\{\| \vec{\zeta}_n \| \geq \varepsilon_3\right\} \leq P\left\{\| \vec{\zeta}_n \| \geq \varepsilon_3, \| \vec{\zeta}_n \| = \| \vec{\Sigma}_n \|\right\} + \delta_0 \leq$$

$$\leq P\left\{\| \vec{\Sigma}_n \| \geq \varepsilon_3\right\} + \delta_0, \quad n > n_0.$$

This inequality implies that

$$P\left\{\| \vec{\zeta}_n \| \geq \varepsilon_3\right\} \leq P\left\{\| \vec{\Sigma}_n \| \geq \varepsilon_3\right\} + \delta_0 \leq P\left\{\|\widetilde{\widetilde{\Phi}}_n G_1'(\vec{x}_n)\| \| \vec{v}_{1n} \| \geq \frac{\varepsilon_3}{3}\right\} +$$

$$+ P\left\{\|\widetilde{\widetilde{\Phi}}_n G_2'(\vec{x}_n)\,\vec{v}_{2n} \| \geq \frac{\varepsilon_3}{3}\right\} + P\left\{\|\widetilde{\widetilde{\Phi}}_n \sqrt{n}\,\widetilde{F}_n(\vec{x}^*)\| \geq \frac{\varepsilon_3}{3}\right\} + \delta_0, \quad n > n_0.$$

$$(2.38)$$

For given $\delta > 0$ and $\varepsilon_1 > 0$

$$P\left\{\|\widetilde{\widetilde{\Phi}}_n G_1'(\vec{x}_n)\| - \|\Phi^{-1}(\vec{x}^*)G_1'(\vec{x}^*)\| < \varepsilon_1\right\} \geq$$

$$\geq P\left\{\|\widetilde{\widetilde{\Phi}}_n G_1'(\vec{x}_n) - \Phi^{-1}(\vec{x}^*)G_1'(\vec{x}^*)\| < \varepsilon_1\right\} \geq 1 - \delta, \quad n > n_1.$$

Suppose that $\varepsilon_2 = \varepsilon = \|\Phi^{-1}(\vec{x}^*)G_1'(\vec{x}^*)\| + \varepsilon\,\varepsilon_1$ where $\varepsilon > 0$ is arbitrary. Then

$$P\left\{\varepsilon\,\|\widetilde{\widetilde{\Phi}}_n G_1'(\vec{x}_n)\| < \varepsilon_2\right\} \geq 1 - \delta, \quad n > n_1.$$

Multiply both sides of inequality in (2.36) on $\|\widetilde{\widetilde{\Phi}}_n G_1'(\vec{x}_n)\|$. Let then in (2.37) $a = \|\widetilde{\widetilde{\Phi}}_n G_1'(\vec{x}_n)\| \, \| \vec{v}_{1n} \|$, $b = \|\widetilde{\widetilde{\Phi}}_n G_1'(\vec{x}_n)\|\,\varepsilon$, $\zeta = \varepsilon_2$. According to (2.36), (2.37)

$$\delta > P\left\{\|\widetilde{\widetilde{\Phi}}_n G_1'(\vec{x}_n)\| \, \| \vec{v}_{1n} \| \geq \varepsilon_2\right\} - P\left\{\|\widetilde{\widetilde{\Phi}}_n G_1'(\vec{x}_n)\|\,\varepsilon \geq \varepsilon_2\right\}, \quad n > n_2.$$

If $\delta = \delta_1/6$ we have

$$P\left\{\|\widetilde{\widetilde{\Phi}}_n G_1'(\vec{x}_n)\| \, \| \vec{v}_{1n} \| \geq \varepsilon_2\right\} < \frac{\delta_1}{3}, \quad n > \max(n_1, n_2). \tag{2.39}$$

It is evident that $\widetilde{\widetilde{\Phi}}_n G_2'(\vec{x}_n)\, \vec{v}_{2n}$ converges in probability to $\vec{0}$. That is why for given ε_2 and $\delta_1 > 0$ there exists such n_3 that

$$P\left\{\|\widetilde{\widetilde{\Phi}}_n G_2'(\vec{x}_n)\, \vec{v}_{2n} \| \geq \varepsilon_2\right\} < \frac{\delta_1}{3}, \quad n > n_3. \tag{2.40}$$

For given $\delta_1 > 0$ we can find such $\varepsilon_4 > 0$ and n_4, that

$$P\left\{\|\widetilde{\widetilde{\Phi}}_n \sqrt{n}\, \widetilde{F}_n(\vec{x}^*)\| \geq \varepsilon_4\right\} < \frac{\delta_1}{3}, \quad n > n_4. \tag{2.41}$$

Changing ε_1 we can obtain $\varepsilon_2 = \varepsilon_4$. Then, when in the right side of (2.38) $\varepsilon_3 = 3\varepsilon_4 = 3\varepsilon_2$, we obtain Lemma 12 from (2.38)- (2.41) putting

$$\varepsilon = \varepsilon_3, \quad \delta = \delta_1 + \delta_0, \quad N_0 = \max(n_0, n_1, n_2, n_3, n_4).$$

Proof of Theorem 14. Consider the quadratic programming problem

$$\widetilde{\varphi}_n(\vec{x}) = \frac{1}{2}\,\vec{x}'\,\Phi(\vec{x}^*)\,\vec{x} + \sqrt{n}\,\widetilde{F}_n'(\vec{x}^*)\,\vec{x} \rightarrow \min, \tag{2.42}$$

$$\vec{\nabla}\, g_i'(\vec{x}^*)\,\vec{x} \leq 0, \quad i \in N_1. \tag{2.43}$$

Denote its solution $\tilde{u}_n$. By Lemma 8 $\tilde{u}_n$ is a continuous function of $\sqrt{n}\, \tilde{F}_n(\vec{x}^{\,*})$:

$$\tilde{u}_n = \varphi\left(\sqrt{n}\, \tilde{F}_n(\vec{x}^{\,*})\right).$$

Because

$$\sqrt{n}\, \tilde{F}_n(\vec{x}^{\,*}) \Longrightarrow \mathcal{N}(\vec{0}, B), \quad n \to \infty$$

we have

$$\varphi\left(\sqrt{n}\, \tilde{F}_n(\vec{x}^{\,*})\right) \Longrightarrow \varphi(\vec{\xi}), \quad n \to \infty.$$

Furthermore, $\varphi(\vec{\xi}) = \vec{\zeta}$ is a solution of (2.34) - (2.35). Then

$$\tilde{u}_n \Longrightarrow \vec{\zeta}, \quad n \to \infty. \tag{2.44}$$

Denote

$$O_n = \left\{\vec{x} : b_i(\vec{x}) \leq 0, \; i = \overline{1, m}\right\}, \quad O = \left\{\vec{x} : \vec{\nabla}\, g_i'(\vec{x}^{\,*})\, \vec{x} \leq 0, \quad i \in N_1\right\}.$$

Because of condition 5 and (2.33) O_n is a convex set. By condition 5 $l_i(\vec{x}) \geq 0$, $\vec{x} \in \mathcal{R}^l$. The inequality

$$\vec{\nabla}\, g_i'(\vec{x}^{\,*})\, \vec{\zeta}_n + l_i(\vec{\zeta}_n) \leq 0, \quad i \in N_1$$

implies that $\vec{\nabla}\, g_i'(\vec{x}^{\,*})\, \vec{\zeta}_n \leq 0$, $i \in N_1$, hence $\vec{\zeta}_n \in O$.

Because $\Phi(\vec{x}^{\,*})$ is positive definite (condition 2), $\tilde{\varphi}_n(\vec{x})$ is a strongly convex function in $\vec{x}$ for fixed u. For $\vec{\zeta}_n \in O$ the following relationship with some constant $\mu > 0$ takes place [64]:

$$\| \vec{\zeta}_n - \tilde{u}_n \|^2 \leq \frac{2}{\mu}\left[\tilde{\varphi}_n(\vec{\zeta}_n) - \tilde{\varphi}_n(\tilde{u}_n)\right].$$

Then for any $\varepsilon > 0$

$$P\left\{\| \vec{\zeta}_n - \tilde{u}_n \|^2 < \varepsilon^2\right\} \geq P\left\{\frac{2}{\mu}\left[\tilde{\varphi}_n(\vec{\zeta}_n) - \tilde{\varphi}_n(\tilde{u}_n)\right] < \varepsilon^2\right\} \geq$$

$$\geq P\left\{|\widetilde{\varphi}_n(\overrightarrow{\zeta}_n) - \varphi_n(\overrightarrow{\zeta}_n)| < \frac{\varepsilon_1}{2}\right\} + P\left\{\varphi_n(\overrightarrow{\zeta}_n) - \widetilde{\varphi}_n(\widetilde{u}_n) < \frac{\varepsilon_1}{2}\right\} - 1, \quad (2.45)$$

where $\varepsilon_1 = \mu\varepsilon^2/2$.

Let us estimate probabilities in the right-hand side of (2.45). From (2.32), (2.42) for any $\varepsilon_2 > 0$, $\delta > 0$

$$P\left\{|\widetilde{\varphi}_n(\overrightarrow{\zeta}_n) - \varphi_n(\overrightarrow{\zeta}_n)| < \varepsilon_2\right\} \geq 1 - P\left\{\|\overrightarrow{\zeta}_n\| \geq \sqrt{b}\right\} -$$

$$-P\left\{\|\Phi(\overrightarrow{x}^{*}) - \widetilde{\Phi}_n\| > 2\varepsilon_2/b\right\} > 1 - \delta, \quad n > n_1, \quad (2.46)$$

where b is some positive number. To obtain (2.46), we used Lemmas 12 and 7.

By Lemma 7 and (2.44)

$$P\left\{|\widetilde{\varphi}_n(\widetilde{u}_n) - \varphi_n(\widetilde{u}_n)| < \varepsilon_2\right\} > 1 - \delta, \quad n > n_2. \quad (2.47)$$

Let us estimate the second addend in (2.45). After some transformations we obtain

$$P\left\{\varphi_n(\overrightarrow{\zeta}_n) - \widetilde{\varphi}_n(\widetilde{u}_n) < \frac{\varepsilon_1}{2}\right\} \geq P\left\{\varphi_n(\overrightarrow{\zeta}_n) - \varphi_n(\widetilde{u}_n) \leq 0\right\} +$$

$$+P\left\{|\varphi_n(\widetilde{u}_n) - \widetilde{\varphi}_n(\widetilde{u}_n)| < \frac{\varepsilon_1}{2}\right\} - 1. \quad (2.48)$$

Consider the first addend in the right-hand side of (2.48). If $\widetilde{\Phi}_n$ is a positively definite matrix then the problem (2.32) - (2.33) has a unique solution, because O_n is a convex set and condition 6 holds. Thus

$$P\left\{\varphi_n(\overrightarrow{\zeta}_n) - \varphi_n(\widetilde{u}_n) \leq 0\right\} \geq P\left\{\widetilde{u}_n \in O_n, \ \wedge_{in} > 0, \ i = \overline{1,l}\right\} =$$

$$= P\left\{b_j(\widetilde{u}_n) \leq 0, \quad j = \overline{1,m}, \quad \wedge_{in} > 0, \quad i = \overline{1,l}\right\},$$

where $\wedge_{in}$ − i-th eigenvalue of $\widetilde{\Phi}_n$. Then

$$P\left\{\varphi_n(\overrightarrow{\zeta}_n) - \varphi_n(\widetilde{u}_n) \leq 0\right\} \geq P\left\{\wedge_{in} > 0, \ i = \overline{1,l}\right\} - m + \sum_{i=1}^{m} P\left\{b_i(\widetilde{u}_n) \leq 0\right\}.$$

$$(2.49)$$

From (2.33), (2.44), the continuity of the functions $\widetilde{g}_i$ and their boundness in $\overrightarrow{x} = \overrightarrow{x}^*$ we have

$$p \lim_{n \to \infty} l_i(\widetilde{u}_n) = 0. \tag{2.50}$$

Using (2.44), (2.50) and relationships

$$\overrightarrow{\nabla} g_i'(\overrightarrow{x}^*) \, \overrightarrow{\zeta} \geq 0, \quad i \in N_1;$$

$$g_i(\overrightarrow{x}^*) < 0, \quad i \in N_2,$$

it is easy to prove that for any $\eta_i > 0$ there exists such n_{3i} that

$$P\left\{ b_i(\widetilde{u}_n) \leq 0 \right\} \geq 1 - \eta_i, \quad n > n_{3i}, \quad i = \overline{1, m}. \tag{2.51}$$

Applying Lemma 9 (part 2) to $\widetilde{\Phi}_n$ and using Lemma 7, for any $\eta > 0$ we have

$$P\left\{ \wedge_{in} > 0, \quad i = \overline{1, l} \right\} > 1 - \frac{\eta}{2}, \quad n > n_5.$$

From the last inequality, (2.51) and (2.49), if

$$\eta = 2 \sum_{i=1}^{m} \eta_i, \quad n_4 = \max_{i=\overline{1,m}} n_{3i},$$

then we have

$$P\left\{ \varphi_n(\overrightarrow{\zeta}_n) - \varphi_n(\widetilde{u}_n) \leq 0 \right\} > 1 - \eta, \quad n > n_6 = \max(n_4, n_5). \tag{2.52}$$

From (2.52) and (2.39) for $\varepsilon_2 = \varepsilon_1/2$ according to (2.38) it follows that

$$P\left\{ \varphi_n(\overrightarrow{\zeta}_n) - \widetilde{\varphi}_n(\widetilde{u}_n) \leq \varepsilon_2 \right\} > 1 - \delta - \eta, \quad n > \max(n_2, n_6). \tag{2.53}$$

Using (2.53) and (2.37) in the right-hand side of (2.36) for $\varepsilon_2 = \varepsilon_1/2$, we obtain

$$P\left\{ \| \overrightarrow{\zeta}_n - \widetilde{u}_n \|^2 < \varepsilon_2 \right\} > 1 - 2\delta - \eta, \quad n > \max(n_1, n_2, n_6).$$

Thus

$$p \lim_{n \to \infty} \| \vec{\zeta}_n - \tilde{u}_n \| = 0. \tag{2.54}$$

This and (2.44) imply Theorem 14.

Now consider the case when $\{\vec{\xi}_i, i \geq 1\}$ is a stationary random vector process with discrete time.

Theorem 15 *Assume that the vector* $\vec{\nabla} f(\vec{x}^*, \vec{\xi}_i)$ *satisfies the following conditions:*

1) *the strong mixing condition with* $\alpha(n) \leq \dfrac{c}{n^{1+\varepsilon}}, \ \varepsilon > 0;$

2) $E\| \vec{\nabla} f(\vec{x}^*, \vec{\xi}_i) \|^{2+\delta} < \infty, \ \varepsilon\delta > 2;$

3) *the spectral density* $h(\lambda)$ *of the vector* $\vec{\nabla} f(\vec{x}^*, \vec{\xi}_i)$ *is a nonsingular matrix at the point* $\lambda = 0;$

4) *conditions* $1 - 6$ *hold.*

Then the vector $\vec{\zeta}_n = \sqrt{n}(\vec{x}_n - \vec{x}^*)$, *which solves the problem (2.32) - (2.33), converges weakly to the random vector* $\vec{\zeta}$ *that solves the problem (2.34) - (2.35), where the vector* $\vec{\xi}$ *is* $\mathcal{N}\left(\vec{0}, 2\pi h(0)\right)$ *– distributed vector.*

The proof is similar to Theorem 14.

For dependent vectors that are difference martingales, we have another version of the central limit theorem.

Theorem 16 *[15]. Let* $\left\{u_n, n \geq 1\right\}$ *be stationary in a strict sense metrically transitive random process and assume that* $E\left\{u_n/\mathcal{F}_{n-1}\right\} = 0$ *where* $\mathcal{F}_n$ *is the* σ-*algebra generated by the random vectors* $u_1, \ldots, u_n$. *Then the distribution of the random variable* $\dfrac{1}{\sqrt{N}} \sum_{k=1}^{N} u_k$ *converges weakly as* $N \to \infty$ *to the* $\mathcal{N}(\vec{0}, R_0)$ *– distribution, where* $R_n = E(u_n u_n'/\mathcal{F}_{n-1}), \ R_0 = \lim_{n \to \infty} R_n.$

Theorem 16 leads to the following proposition.

Theorem 17 *Assume that* $\vec{\nabla} f(\vec{x}^{*}, \vec{\xi}_{i})$ *satisfies the following conditions:*

1) $E[\vec{\nabla} f(\vec{x}^{*}, \xi_{i}/G_{i-1})] = \vec{0}$, *where* G_{i} *is the* σ-*algebra generated by the vectors* $\vec{\nabla} f(\vec{x}^{*}, \vec{\xi}_{1}), \ldots, \vec{\nabla} f(\vec{x}^{*}, \vec{\xi}_{i})$;

2) conditions $1 - 6$ *hold.*

Then the vector $\vec{\zeta}_{n} = \sqrt{n}(\vec{x}_{n} - \vec{x}^{*})$, *which solves the problem (2.32) - (2.33) converges weakly to the random vector* $\vec{\zeta}$ *that solves the problem (2.34) - (2.35), where* $\vec{\xi}$ *is the* $\mathcal{N}(\vec{0}, 2\pi R_{0})$ *- distributed vector with the matrix* R_{0} *of the form*

$$R_{0} = \left(E \frac{\partial f(\vec{x}^{*}, \vec{\xi}_{1})}{\partial x_{i}} \frac{\partial f(\vec{x}^{*}, \vec{\xi}_{1})}{\partial x_{j}} \right)_{i,j=1}^{l}.$$

Let now $\left\{ \vec{\xi}(t), t \in \mathcal{R} \right\}$ be a strictly stationary ergodic random process and

$$F_{T}(\vec{x}) = \frac{1}{T} \int_{0}^{T} f\left(\vec{x}, \vec{\xi}(t) \right) dt,$$

$$\vec{x}_{T} \in Arg \min_{x \in I} F_{T}(\vec{x}).$$

Theorem 18 *Let conditions* $1 - 6$ *and following ones be satisfied:*

1) the strong mixing condition with $\alpha(\tau) \leq \frac{c}{\tau^{1+\varepsilon}}$, $\varepsilon > 0$;

2) $E \| \vec{\nabla} f\left(\vec{x}^{*}, \vec{\xi}(0) \right) \|^{2+\delta} < \infty$, $\varepsilon\delta > 2$;

3) the spectral density $h(\lambda)$ *of the vector process* $\vec{\nabla} f\left(\vec{x}^{*}, \vec{\xi}(t) \right)$ *is a nonsingular matrix at the point* $\lambda = 0$.

Then the vector $\vec{\zeta}_T = \sqrt{T}\,(\vec{x}_T - \vec{x}^{\,*})$ converges weakly as $T \to \infty$ to the random vector $\vec{\zeta}$ which is the solution to the problem (2.34) - (2.35).

The proof is similar to Theorem 14.

2.5 NONSTATIONARY EMPIRICAL ESTIMATES

Consider now conditions of consistency and asymptotic distribution of empirical estimates for the case when f also depends on time.

Let $\left\{\vec{\xi}_n,\, n \geq 0\right\}$ be random process with discrete time defined on the prob-

abilistic space $(\Omega, \mathcal{F}, P)$, $\vec{\xi}_n \in \mathcal{R}^m$, $m \geq 1$, I is a closed subset in $\mathcal{R}^p$, $p \geq 1$, $\|\cdot\|$ is the norm in $\mathcal{R}^p$, the function

$$\left\{f(i, \vec{x}, \vec{y}) \,:\, \mathcal{N} \times I \times \mathcal{R}^m \to \mathcal{R}_+, \quad \mathcal{R}_+ = [0, \infty)\right\}$$

is separable semi-continuous on the second parameter and measurable on the third one.

The problem is to minimize the functional

$$F_n(\vec{x}) = \frac{1}{n} \sum_{i=1}^{n} f(i, \vec{x}, \vec{\xi}_i), \quad \vec{x} \in I.$$

We will need the following statement.

Theorem 19 *Suppose that in Theorem 7 functions* $\left\{Q_T(s),\, T \in \mathcal{R}(\mathcal{N})\right\}$ *and* $\Phi(s; s_0)$ *are separable, continuous on sets* $\|s - s_0\| > \delta$ *for any* $\delta > 0$, *semi-continuous in* s_0, *and conditions 2- 4 are satisfied. Then the element* s_T *can be chosen as measurable in* ω *and*

$$P\left\{\lim_{T \to \infty} \|s_T - s_0\| = 0\right\} = 1.$$

The proof is similar to Theorem 7 [15].

Theorem 20 *Let the stochastic function $f(i, \vec{x}, \vec{\xi}_i)$ satisfy the following conditions:*

1. *For every $\vec{x} \in I$ there exists function $F(\vec{x})$ such that*

$$F(\vec{x}) = \lim_{n \to \infty} E\, F_n(\vec{x})$$

   and the point $\vec{x}^* \in I$ such that

$$F(\vec{x}^*) < F(\vec{x}), \quad \vec{x} \neq \vec{x}^* .$$

2. *For any $\delta > 0$ the function $f(i, \vec{x}, \vec{y})$ is continuous in the second parameter in the region $\left\{ \vec{x} : \| \vec{x} - \vec{x}^* \| \geq \delta \right\}$.*

3. *If I is unbounded set then*

$$f(i, \vec{x}, \vec{y}) \to \infty, \quad \| \vec{x} \| \to \infty$$

   under fixed i and $\vec{y}$.

4. *There exists the function $c(\gamma) > 0$ such that $c(\gamma) \to 0$ as $\gamma \to 0$, and for any $\delta > 0$ there exists the $\gamma_0 > 0$ such that for any element $\vec{x}_1 \in I$, $0 < \gamma < \gamma_0$*

$$\varlimsup_{n \to \infty} \sum_{i=1}^{n} \sup_{\substack{\|\vec{x} - \vec{x}_1\| < \gamma, \\ \|\vec{x} - \vec{x}^*\| > \delta}} \left| f(i, \vec{x}, \vec{\xi}_i) - f(i, \vec{x}_1, \vec{\xi}_i) \right| < c(\gamma).$$

5. *The function $f(i, \vec{x}, \vec{\xi}_i)$ satisfies the strong mixing condition with $\alpha(j) \leq \dfrac{c}{1 + j^{1+\varepsilon}}$, $\varepsilon > 0$.*

6. *$E f(i, \vec{x}, \vec{\xi}_i)^{2+\delta} < \infty$, where $\varepsilon \delta > 2$.*

Denote $\vec{x}_n = \arg\min_{\vec{x} \in I} F_n(\vec{x})$. Then

$$P\left\{ \lim_{n \to \infty} \| \vec{x}_n - \vec{x}^* \| = 0 \right\} = 1.$$

Proof. It is easy to see that under condition 3 the element $\vec{x}_n$ belongs to the ball $K = \left\{ \vec{x} : \| \vec{x} \| \leq r \right\}$ with probability 1. We need to verify whether conditions of Theorem 19 hold. We will prove that for any $\vec{x} \in K$

$$P\left\{ \lim_{n \to \infty} F_n(\vec{x}) = F(\vec{x}) \right\} = 1.$$

Under conditions 5, 6 and taking into account [54], [63], it is easy to see that

$$\left| Ef(i, \vec{x}, \vec{\xi}_i) \, f(j, \vec{x}, \vec{\xi}_j) - Ef(i, \vec{x}, \vec{\xi}_i) \, Ef(j, \vec{x}, \vec{\xi}_j) \right| \leq$$

$$\leq \frac{c}{1 + |i - j|^{1+\varepsilon'}}, \quad \varepsilon' > 0.$$

Denote $\eta_n(\vec{x}) = F_n(\vec{x}) - EF_n(\vec{x})$ and estimate $E\eta_n^2(\vec{x})$:

$$E\eta_n^2(\vec{x}) = E\left\{ \frac{1}{n} \sum_{i=1}^{n} f(i, \vec{x}, \vec{\xi}_i) - E\left(\frac{1}{n} \sum_{i=1}^{n} f(i, \vec{x}, \vec{\xi}_i) \right) \right\}^2 =$$

$$= E\left\{ \frac{1}{n^2} \sum_{i=1}^{n} \sum_{j=1}^{n} \left[f(i, \vec{x}, \vec{\xi}_i) - Ef(i, \vec{x}, \vec{\xi}_i) \right] \left[f(j, \vec{x}, \vec{\xi}_j) - \right. \right.$$

$$\left. \left. - Ef(j, \vec{x}, \vec{\xi}_j) \right] \right\} = \frac{1}{n^2} \sum_{i=1}^{n} \sum_{j=1}^{n} Ey_i y_j,$$

where $y_i = f(i, \vec{x}, \vec{\xi}_i) - Ef(i, \vec{x}, \vec{\xi}_i)$.

Under conditions 5 and 6 we obtain

$$Ey_i y_j \leq \frac{c}{1 + |i - j|^{1+\varepsilon'}}, \quad \varepsilon' > 0.$$

That is why

$$\frac{1}{n^2} \sum_{i=1}^{n} \sum_{j=1}^{n} Ey_i y_j \leq \frac{c}{n^2} \sum_{i=1}^{n} \sum_{j=1}^{n} \frac{1}{1 + |i - j|^{1+\varepsilon'}} \leq \frac{c}{n}.$$

Let $n = m^2$. Then by Borel-Cantelly lemma we have

$$P\left\{\lim_{m\to\infty} \eta_{m^2} = 0\right\} = 1.$$

Let

$$\zeta_m = \sup_{m^2 \le n \le (m+1)^2} |\eta_n - \eta_{m^2}|.$$

For $m^2 \le n \le (m+1)^2$ the following inequality holds:

$$|\eta_n| \le |\eta_{m^2}| + \sup_{m^2 \le n \le (m+1)^2} |\eta_n - \eta_{m^2}|.$$

For ζ_m we have

$$E(\zeta_m)^2 = E \sup_{m^2 \le n \le (m+1)^2} |\eta_n - \eta_{m^2}| \le$$

$$\le \sup_{m^2 \le n \le (m+1)^2} \left| \frac{1}{m^2} \sum_{i=1}^{n} \sum_{j=1}^{n} E y_i\, y_j \right| \le$$

$$\le \frac{1}{m^2} \sum_{i=m^2+1}^{(m+1)^2} \sum_{j=m^2+1}^{(m+1)^2} E |y_i\, y_j| \le$$

$$\le c \left[\frac{(m+1)^2 - m^2 - 1}{m^2} \right]^2 = \frac{2c}{m^2}.$$

Therefore by Borel-Cantelli lemma

$$P\left\{\lim_{n\to\infty} \zeta_n = 0\right\} = 1.$$

Then, $P\left\{\lim_{n\to\infty} \eta_n = 0\right\} = 1.$

Thus, $P\left\{\lim_{n\to\infty} F_n(\vec{x}) = F(\vec{x})\right\} = 1.$

In fact, condition 3 of Theorem 19 holds. Let us varify condition 4:

$$\sup_{\{\vec{x}\, :\, \|\vec{x} - \vec{x}_1\| < \gamma,\, \|\vec{x} - \vec{x}^*\| > \varepsilon\}} \left| F_n(\vec{x}) - F_n(\vec{x}_1) \right| =$$

$$
= \sup_{\{\vec{x}\,:\,\|\vec{x}-\vec{x}_1\|<\gamma,\,\|\vec{x}-\vec{x}^*\|>\varepsilon\}} \left| \frac{1}{n}\sum_{i=1}^{n} f(i,\vec{x},\vec{\xi}_i) - \frac{1}{n}\sum_{i=1}^{n} f(i,\vec{x}_1,\vec{\xi}_i) \right| \le
$$

$$
\le \frac{1}{n}\sum_{i=1}^{n} \sup_{\{\vec{x}\,:\,\|\vec{x}-\vec{x}_1\|<\gamma,\,\|\vec{x}-\vec{x}^*\|>\varepsilon\}} \left| f(i,\vec{x},\vec{\xi}_i) - f(i,\vec{x}_1,\vec{\xi}_i) \right|.
$$

Under condition 1

$$
\lim_{n\to\infty} \frac{1}{n}\sum_{i=1}^{n} \sup_{\{\|\vec{x}-\vec{x}_1\|<\gamma,\,\|\vec{x}-\vec{x}^*\|>\varepsilon\}} \left| f(i,\vec{x},\vec{\xi}_i) - f(i,\vec{x}_1,\vec{\xi}_i) \right| \le
$$

$$
\le \lim_{n\to\infty} \frac{1}{n}\sum_{i=1}^{n} \sup_{\{\vec{x}\,:\,\|\vec{x}-\vec{x}_1\|<\gamma,\,\|\vec{x}-\vec{x}^*\|>\varepsilon\}} \left| f(i,\vec{x},\vec{\xi}_i) - f(i,\vec{x}_1,\vec{\xi}_i) - \right.
$$

$$
- \left. E\left[f(i,\vec{x},\vec{\xi}_i) - f(i,\vec{x}_1,\vec{\xi}_i) \right] \right| +
$$

$$
+ \lim_{n\to\infty} \frac{1}{n}\sum_{i=1}^{n} \sup_{\{\vec{x}\,:\,\|\vec{x}-\vec{x}_1\|<\gamma,\,\|\vec{x}-\vec{x}^*\|>\varepsilon\}} E\left| f(i,\vec{x},\vec{\xi}_i) - f(i,\vec{x}_1,\vec{\xi}_i) \right|.
$$

$$(2.55)$$

Similarly to the previous argumentation for condition 3 of Theorem 19 it is easy to show that the first item in (2.55) converges to 0 with probability 1. For the second item of (2.55) we have

$$
\lim_{n\to\infty} \frac{1}{n}\sum_{i=1}^{n} \sup_{\{\vec{x}\,:\,\|\vec{x}-\vec{x}_1\|<\gamma,\,\|\vec{x}-\vec{x}^*\|>\varepsilon\}} E\left| f(i,\vec{x},\vec{\xi}_i) - f(i,\vec{x}_1,\vec{\xi}_i) \right| <
$$

$$
< \lim_{n\to\infty} \frac{1}{n}\sum_{i=1}^{n} \sup_{\{\vec{x}\,:\,\|\vec{x}-\vec{x}_1\|<\gamma,\,\|\vec{x}-\vec{x}^*\|>\varepsilon\}} \left| f(i,\vec{x},\vec{\xi}_i) - f(i,\vec{x}_1,\vec{\xi}_i) \right|.
$$

Denote

$$
c(\gamma) = 2 \lim_{n\to\infty} \frac{1}{n}\sum_{i=1}^{n} E \sup_{\{\vec{x}\,:\,\|\vec{x}-\vec{x}_1\|<\gamma,\,\|\vec{x}-\vec{x}^*\|>\varepsilon\}} \left| f(i,\vec{x},\vec{\xi}_i) - f(i,\vec{x}_1,\vec{\xi}_i) \right|.
$$

Taking into account the continuity of the function $f(i, \vec{x}, \vec{y})$ in the second parameter in the region $\left\{\vec{x} : \| \vec{x} - \vec{x}^* \| \geq \varepsilon\right\}$ we obtain that $c(\gamma) \to 0$ at $\gamma \to 0$. Then condition 4 of Theorem 19 holds. The theorem is proved.

Consider a random process with a continuous parameter.

Let $\left\{\vec{\xi}(t), t \geq 0\right\}$ be a random process with continuous time defined on the probabilistic space $(\Omega, \mathcal{F}, P)$, $\vec{\xi}(t) \in \mathcal{R}^m$, $m \geq 1$, I is the closed subset in $\mathcal{R}^p$, $\left\{f(t, \vec{x}, \vec{y}) : \mathcal{R}_+ \times I \times \mathcal{R}^p \to \mathcal{R}_+\right\}$ is continuous in all parameters function. The problem is to minimize the functional

$$F_T(\vec{x}) = \frac{1}{T} \int\limits_0^T f\left(t, \vec{x}, \vec{\xi}(t)\right) dt, \quad \vec{x} \in I.$$

Theorem 21 *Let the random function $f\left(t, \vec{x}, \vec{\xi}(t)\right)$ satisfy conditions:*

1. *For any $\vec{x} \in I$ there exists the function $F(\vec{x})$ such that*

$$F(\vec{x}) = \lim_{T \to \infty} E F_T(\vec{x})$$

and the point $\vec{x}^* \in I$ such that

$$F(\vec{x}^*) < F(\vec{x}), \quad \vec{x} \neq \vec{x}^*.$$

2. *If I is unbounded set then*

$$f(t, \vec{x}, \vec{y}) \to \infty, \quad \| \vec{x} \| \to \infty$$

for any fixed t and $\vec{y}$.

3. *There exists the function $c(\gamma) > 0$ such that $c(\gamma) \to 0$, $\gamma \to 0$, and for any $\delta > 0$ there exists $\gamma_0 > 0$ such that for any element $\vec{x}_1 \in I$, $0 < \gamma < \gamma_0$*

$$\overline{\lim_{T \to \infty}} \frac{1}{T} \int\limits_0^T E \sup_{\substack{\| \vec{x} - \vec{x}_1 \| < \gamma, \\ \| \vec{x} - \vec{x}^* \| > \delta}} \left| f\left(t, \vec{x}, \vec{\xi}(t)\right) - f\left(t, \vec{x}_1, \vec{\xi}(t)\right) \right| dt < c(\gamma).$$

4. The process $\vec{\xi}(t)$ satisfies the strong mixing condition with the coefficient $\alpha(\tau) = O(\tau^{-1-\varepsilon})$, $\varepsilon > 0$.

5. $\sup\limits_{t \geq 0} E\left[f\left(t, \vec{x}, \vec{\xi}(t)\right)^{2+\delta}\right] < \infty$, where $\varepsilon\delta > 2$, $\| \vec{x} \| < \infty$.

Denote

$$\vec{x}_T = \arg\min_{\vec{x} \in I} F_T(\vec{x}).$$

Then

$$P\left\{ \lim_{T \to \infty} \| \vec{x}_T - \vec{x}^* \| = 0 \right\} = 1.$$

Proof. Under conditions of the theorem there exists such a number $r > 0$ that the element $\vec{x}_T$ belongs to the ball

$$K = \left\{ \vec{x} : \| \vec{x} \| \leq r \right\}$$

with probability 1. Using this evident fact and Theorem 19 we have to prove that for any $\vec{x} \in K$

$$P\left\{ \lim_{T \to \infty} F_T(\vec{x}) = F(\vec{x}) \right\} = 1.$$

Because of conditions 4, 5 and [61], [63] we obtain that

$$\left| E f\left(t, \vec{x}, \vec{\xi}(t)\right) f\left(t+\tau, \vec{x}, \vec{\xi}(t+\tau)\right) - E f\left(t, \vec{x}, \vec{\xi}(t)\right) \times \right.$$

$$\left. \times E f(t+\tau, \vec{x}, \vec{\xi}(t+\tau)) \right| \leq \frac{c}{1 + \tau^{1+\varepsilon_1}},$$

where $\varepsilon_1 > 0$. Denote

$$\eta_T(\vec{x}) = F_T(\vec{x}) - E F_T(\vec{x})$$

and estimate $E \eta_T^2(\vec{x})$:

$$E \eta_T^2(\vec{x}) = E\left[\frac{1}{T} \int_0^T f\left(t, \vec{x}, \vec{\xi}(t)\right) dt - E \frac{1}{T} \int_0^T f\left(t, \vec{x}, \vec{\xi}(t)\right) dt \right]^2 =$$

$$= E \frac{1}{T^2} \int\limits_0^T \int\limits_0^T \left[f\left(t_1, \vec{x}, \vec{\xi}(t_1)\right) - E f\left(t_1, \vec{x}, \vec{\xi}(t_1)\right) \right] \times$$

$$\times \left[f\left(t_2, \vec{x}, \vec{\xi}(t_2)\right) - E f\left(t_2, \vec{x}, \vec{\xi}(t_2)\right) \right] dt_1 dt_2 =$$

$$= \frac{1}{T^2} \int\limits_0^T \int\limits_0^T E y_1 y_2 dt_1 dt_2,$$

where $y_i = f\left(t_i, \vec{x}, \vec{\xi}(t_i)\right) - E f\left(t_i, \vec{x}, \vec{\xi}(t_i)\right)$.

From conditions 4 and 5 we have

$$\left| E y_1 y_2 \right| \leq \frac{c}{1 + |t_1 - t_2|^{1+\varepsilon_1}}, \quad \varepsilon_1 > 0.$$

That is why

$$\frac{1}{T^2} \int\limits_0^T \int\limits_0^T E y_1 y_2 \, dt_1 dt_2 \leq \frac{c}{T}.$$

Choosing $T = T(n) = n^2$ we obtain by Borel-Cantelli Lemma that

$$P\left\{ \lim_{n \to \infty} \eta_{T(n)}(\vec{x}) = 0 \right\} = 1.$$

Denote

$$\zeta_n = \sup_{T(n) \leq T \leq T(n+1)} \left| \eta_T(\vec{x}) - \eta_{T(n)}(\vec{x}) \right|.$$

It is evident that

$$\zeta_n \leq \frac{T(n+1) - T(n)}{T^2(n)} \left| \eta_{T(n)}(\vec{x}) \right| +$$

$$+ \frac{1}{T^2(n)} \max_{T(n) \leq T \leq T(n+1)} \left| \int\limits_{T(n)}^T \left[f\left(t, \vec{x}, \vec{\xi}(t)\right) - E f\left(t, \vec{x}, \vec{\xi}(t)\right) \right] dt \right|.$$

Then, the first addend of the last relationship converges to zero with probability 1. Using conditions 4, 5 and Borel-Cantelli Lemma we can prove similarly that the second addend converges to zero as $n \to \infty$ with probability 1. Thus,

$$P\left\{\lim_{T \to \infty} F_T(\vec{x}) = F(\vec{x})\right\} = 1.$$

To use Theorem 19 we have to verify the following condition: for any $\delta > 0$ there exists $\gamma_0 > 0$ and the function $c(\gamma)$, $\gamma > 0$ such that $c(\gamma) \to 0$, $\gamma \to 0$ and for any element $\vec{x}_1 \in K$ and any $0 < \gamma < \gamma_0$

$$P\left\{\overline{\lim_{T \to \infty}} \sup_{\|\vec{x} - \vec{x}_1\| < \gamma, \|\vec{x} - \vec{x}^*\| > \delta} \left|F_T(\vec{x}) - F_T(\vec{x}_1)\right| < c(\gamma)\right\} = 1.$$

Because of conditions $3 - 5$ we have

$$\overline{\lim_{T \to \infty}} \sup_{\|\vec{x} - \vec{x}_1\| < \gamma, \|\vec{x} - \vec{x}^*\| > \delta} \left|F_T(\vec{x}) - F_T(\vec{x}_1)\right| \le$$

$$\le \overline{\lim_{T \to \infty}} \frac{1}{T} \int_0^T E \sup_{\|\vec{x} - \vec{x}_1\| < \gamma, \|\vec{x} - \vec{x}^*\| > \delta} \left|f\left(t, \vec{x}, \vec{\xi}\,(t)\right) - \right.$$

$$\left. - f\left(t, \vec{x}_1, \vec{\xi}\,(t)\right)\right| dt = \frac{1}{2} c(\gamma).$$

Then

$$P\left\{\overline{\lim_{T \to \infty}} \sup_{\|\vec{x} - \vec{x}_1\| < \gamma, \|\vec{x} - \vec{x}^*\| > \delta} \left|F_T(\vec{x}) - F_T(\vec{x}_1)\right| < c(\gamma)\right\} = 1.$$

The theorem is proved.

Further we shall study the asymptotic distribution of the estimates. As the destriction set I we take following one:

$$I = \left\{\vec{x} : \vec{g}\,(\vec{x}) = \left(g_1(\vec{x}), \dots, g_m(\vec{x})\right) \le \vec{0}\right\}.$$

Let ξ_i, $i \ge 1$ be independent identically distributed random variables, $E\,\xi_i = 0$, $E\,\xi_i^2 = \sigma^2$. As for functionals (2.26) we suppose conditions $1 - 6$ to be fulfilled, and besides the following condition:

7. *There exist limits*

$$F_{kj}(\vec{x}) = \lim_{n \to \infty} \frac{1}{n} \sum_{i=1}^{n} E \frac{\partial f(i, \vec{x}, \xi_i)}{\partial x_k} \frac{\partial f(i, \vec{x}, \xi_i)}{\partial x_j},$$

$$\Phi_{kj}^{\gamma}(\vec{x}^*) = \overline{\lim_{T \to \infty}} \frac{1}{n} \sum_{i=1}^{n} E \sup_{\vec{x} \,:\, \|\vec{x} - \vec{x}^*\| < \gamma} \left| \frac{\partial^2 f(i, \vec{x}, \xi_i)}{\partial x_k \partial x_j} \right|.$$

We have

$$F_n(\vec{x}) = \frac{1}{n} \sum_{i=1}^{n} f(i, \vec{x}, \xi_i),$$

where f is continuous in the second argument and measurable in the third one.

Theorem 22 *Let conditions $1 - 7$ be satisfied. Then the random vector $\vec{\zeta}_n = \sqrt{n}(\vec{x}_n - \vec{x}^*)$, which is a solution of problem (2.32) - (2.33), converges in distribution to the random vector $\vec{\zeta}$ - the solution of problem (2.34) - (2.35), where $\vec{\xi}$ is $\mathcal{N}(\vec{0}, \sigma^2 \widetilde{F})$ - distributed, $\widetilde{F}$ is the matrix with elements $F_{kj}(\vec{x}^*)$.*

The proof is similar to Theorem 14.

Let now $\left\{ \xi_i, i \geq 1 \right\}$ be a stationary in a strict sense random process.

Theorem 23 *Suppose that the process $\vec{\nabla} f(i, \vec{x}^*, \vec{\xi}_i)$ satisfies the strong mixing condition with the coefficient $\alpha(n) \leq \frac{c}{n^{1+\varepsilon}}$, conditions $1 - 7$ are fulfilled and*

$$E \| \vec{\nabla} f(i, \vec{x}^*, \vec{\xi}_i) \|^{2+\delta} < \infty, \quad \varepsilon\delta > 2.$$

Then the vector $\vec{\zeta}_n = \sqrt{n}(\vec{x}_n - \vec{x}^*)$, converges weakly to $\vec{\zeta}$ - the solution to problem (2.34) - (2.35), where $\vec{\xi}$ is $\mathcal{N}(\vec{0}, \sigma^2 \widetilde{F})$ - distributed.

Theorem 23 may be proved in the same way as Theorem 14 was.

3

PARAMETRIC REGRESSION MODELS

In this chapter different cases of parametric regression models are considered.

At first we investigate the model where a sequence of independent or dependent weakly nonidentically distributed Gaussian random variables is observed. It is supposed that the monotone functions of the mathematical expectations and variances are represented as the linear combinations of unknown parameters. The problem is to estimate the parameters by observations of the random variables and the coefficients in the representations. The modified least squares estimate is considered, and the consistency of the estimate is proved.

Then the properties of the estimator of the linear regression parameters on Gaussian random fields are considered. The least squares estimates are invest-igated, the theorems about the consistency and asymptotic normality of the estimates are proved. It is shown that the estimates have the moments of any order if the observation interval is large enough, and that the moments converge to the corresponding moments of Gaussian distribution.

Further two variants of a nonlinear regression model are investigated. The asymptotic properties of the least modules and least squares estimates are studied. The identification problem is represented as a special case of the stochastic programming problem considered above.

Then nonstationary regression model for the random field observed in the circle is studied. It is proved that the least squares estimates of the unknown para-meters are consistent and asymptotically normal.

At the end of the chapter the Gaussian regression model for quasistationary random processes is considered. The consistency of the least squares estimates is proved.

3.1 ESTIMATES OF THE PARAMETERS FOR GAUSSIAN REGRESSION MODELS WITH DISCRETE TIME

First we shall consider independent nonidentically distributed Gaussian random variables.

Let a sequence of independent Gaussian random variables $y_i = \mathcal{N}(\mu_i, \sigma_i^2)$ be observed, where μ_i and σ_i^2 are the mathematical expectation and variance of y_i, respectively. Suppose that functions $g_1(t)$ and $g_2(t)$ are continuous, strictly monotone, and related to μ_i and $\sigma_i(t)$ as follows:

$$g_1(\mu_i) = \eta_i = \sum_{j=1}^{k} x_{ij}\beta_j^0, \quad g_2(\sigma_i^2) = \zeta_i = \sum_{j=1}^{l} z_{ij}\gamma_j^0.$$

The parameters $\vec{\beta}^0 = (\beta_1^0, \ldots, \beta_k^0) \subset B \subset \mathcal{R}^k$ and $\vec{\gamma}^0 = (\gamma_1^0, \ldots, \gamma_k^0) \subset \Gamma \subset \mathcal{R}^l$ are unknown, where B and Γ are compact sets. The vectors $\vec{x}_i = (x_{i1}, \ldots, x_{ik})$ and $\vec{z}_i = (z_{i1}, \ldots, z_{il})$ are unknown and nonrandom. The aim is to estimate unknown parameters $\vec{\beta}^0$ and $\vec{\gamma}^0$ from the observations $(y_i, \vec{x}_i, \vec{z}_i)$. Further we assume that the sets of points $\vec{x}_i$ and $\vec{z}_i$ satisfy the following conditions [104]: there exist r_1 and r_2 such that

$$\lambda_1(A) = \lim_{n \to \infty} \frac{1}{n}\chi_{[\vec{x}_i \in A]} > 0, \tag{3.1}$$

$$\lambda_2(S) = \lim_{n \to \infty} \frac{1}{n}\chi_{[\vec{z}_i \in S]} > 0, \tag{3.2}$$

for an arbitrary open $A \in B_{r_1}(\vec{0})$ and $S \in B_{r_2}(\vec{0})$, where $B_{r_1}(\vec{0})$ and $B_{r_2}(\vec{0})$ are open balls in the spaces $\mathcal{R}^k$ and $\mathcal{R}^l$ with origins at $\vec{0}$ and radii r_1 and r_2, respectively. Here χ_c is the indicator function of a set C. We also assume that $\|\vec{x}_i\| \leq C_x$ and $\|\vec{z}_i\| \leq C_z$.

To estimate $\vec{\beta}^0$ and $\vec{\gamma}^0$ we consider the functional

$$F_n(\vec{\beta}, \vec{\gamma}) = \frac{1}{n} \sum_{i=1}^{n} \frac{\left[y_i - g_1^{-1} \left(\sum_{j=1}^{k} x_{ij} \beta_j \right) \right]^2}{g_2^{-1} \left(\sum_{j=1}^{l} z_{ij} \gamma_j \right)} + $$

$$+ \frac{1}{n} \sum_{i=1}^{n} \log 2\pi \, g_2^{-1} \left(\sum_{j=1}^{l} z_{ij} \gamma_j \right).$$

The functions $g_1^{-1}(t)$ and $g_2^{-1}(t)$ exist under the above conditions on $g_1(t)$ and $g_2(t)$.

We take the values of $\vec{\beta}^n$ and $\vec{\gamma}^n$ that minimize the functional $F_n(\vec{\beta}, \vec{\gamma})$ as estimates for the unknown parameters $\vec{\beta}^0$ and $\vec{\gamma}^0$.

Definition 1 [104], p.44. *Suppose that B is a subset of some Banach space, $(B, \mathcal{B})$ is the measurable space, $(\Omega, \mathcal{G}, P)$ is the probabilistic space, $I : B \times \Omega \to \mathcal{R}$ is the function. The function I is called a normal integrant on $B \times \Omega$ if it is measurable as a function of $(\vec{\beta}, \omega)$ and lower semicontinuous with respect to $\vec{\beta} \in B$ (i.e. the sets $\{\vec{\beta} : I(\vec{\beta}, \omega) \leq c\}$ are closed) for almost every $\omega \in \Omega$.*

If the function $I : B \times \Omega \to \mathcal{R}$ is continuous in the first argument and measurable in the second one, then it is a normal integrant on $B \times \Omega$, where $\mathcal{B}$ is the σ-algebra of Borel subsets of B.

We will use the following assertion.

Lemma 13 [104], p.44. *Assume that*

$$\left\{ I_n : B \times \Omega \to \mathcal{R}, \quad n \in \mathcal{N} \right\}$$

is a sequence of normal integrants; $\vec{\beta}_n : \Omega \to B$ is a measurable mapping and

$$\vec{\beta}_n(\omega) = \arg \min_{\vec{\beta} \in B} I_n(\vec{\beta}, \omega).$$

Let the set B be compact and the following conditions be satisfied:

1) for some fixed $\vec{\beta}^*\in B$ and any $\vec{\beta}\neq\vec{\beta}^*$

$$P\left\{\lim_{n\to\infty}\tilde{I}_n(\vec{\beta})>0\right\}=1,$$

where $\tilde{I}_n(\vec{\beta})=I_n(\vec{\beta})-I_n(\vec{\beta}^*)$;

2) for each $\vec{\beta}_1\in B$

$$P\left\{\lim_{\delta\to 0}\overline{\lim_{n\to\infty}}\sup_{\vec{\beta}:\|\vec{\beta}-\vec{\beta}_1\|\leq\delta}|I_n(\beta)-I_n(\beta_1)|=0\right\}=1.$$

Then

$$P\left\{\|\vec{\beta}_n-\vec{\beta}^*\|\to 0,\quad n\to\infty\right\}=1.$$

The following assertion holds.

Theorem 24 *Let conditions (3.1) and (3.2) be satisfied and*

$$\inf_{t\in\mathcal{R}}g_2^{-1}(t)=\alpha>0, \tag{3.3}$$

$$\sup_{t\in\mathcal{R}}g_1^{-1}(t)\leq C_1<\infty, \tag{3.4}$$

$$\sup_{t\in\mathcal{R}}g_2^{-1}(t)\leq C_2<\infty. \tag{3.5}$$

Then

$$P\left\{\lim_{n\to\infty}\left(\|\vec{\beta}^n-\vec{\beta}^0\|+\|\vec{\gamma}^n-\vec{\gamma}^0\|\right)=0\right\}=1.$$

Proof. To estimate $I_n(\vec{\beta}, \vec{\gamma}) = F_n(\vec{\beta}, \vec{\gamma}) - F_n(\vec{\beta}^0, \vec{\gamma}^0)$, we check conditions 1, 2 of Lemma 13. At first,

$$
\lim_{n\to\infty} I_n(\vec{\beta}, \vec{\gamma}) = \lim_{n\to\infty} \left\{ \frac{1}{n} \sum_{i=1}^{n} \frac{1}{g_2^{-1}\left(\sum\limits_{j=1}^{l} z_{ij}\gamma_j\right)} \left[\left(g_1^{-1}\left(\sum_{j=1}^{k} x_{ij}\beta_j\right)\right)^2 - \right.\right.
$$

$$
\left. - \left(g_1^{-1}\left(\sum_{j=1}^{k} x_{ij}\beta_j^0\right)\right)^2 - 2y_i \left(g_1^{-1}\left(\sum_{j=1}^{k} x_{ij}\beta_j\right) - g_1^{-1}\left(\sum_{j=1}^{k} x_{ij}\beta_j^0\right)\right) \right] +
$$

$$
+ \frac{1}{n} \sum_{i=1}^{n} \left[y_i - g_1^{-1}\left(\sum_{j=1}^{k} x_{ij}\beta_j\right) \right]^2 \left[\frac{1}{g_2^{-1}\left(\sum\limits_{j=1}^{l} z_{ij}\gamma_j\right)} - \frac{1}{g_2^{-1}\left(\sum\limits_{j=1}^{l} z_{ij}\gamma_j^0\right)} \right] +
$$

$$
\left. + \frac{1}{n} \sum_{i=1}^{n} \log \frac{g_2^{-1}\left(\sum\limits_{j=1}^{l} z_{ij}\gamma_j\right)}{g_2^{-1}\left(\sum\limits_{j=1}^{l} z_{ij}\gamma_j^0\right)} \right\}.
$$

The strong law of large numbers implies that

$$
\lim_{n\to\infty} \left[-\frac{1}{n} \sum_{i=1}^{n} y_i \left(g_1^{-1}\left(\sum_{j=1}^{k} x_{ij}\beta_j\right) - g_1^{-1}\left(\sum_{j=1}^{k} x_{ij}\beta_j^0\right) \right) \right] =
$$

$$
= \lim_{n\to\infty} \left[-\frac{1}{n} \sum_{i=1}^{n} g_1^{-1}\left(\sum_{j=1}^{k} x_{ij}\beta_j^0\right) \left[g_1^{-1}\left(\sum_{j=1}^{k} x_{ij}\beta_j\right) - \right.\right.
$$

$$
\left.\left. - g_1^{-1}\left(\sum_{j=1}^{k} x_{ij}\beta_j^0\right) \right] \right].
$$

Thus

$$
\lim_{n\to\infty} I_n(\vec{\beta}, \vec{\gamma}) = \lim_{n\to\infty} \left\{ \frac{1}{n} \sum_{i=1}^{n} \frac{1}{g_2^{-1}\left(\sum_{j=1}^{l} z_{ij}\gamma_j\right)} \times \right.
$$

$$
\times \left[\left(g_1^{-1}\left(\sum_{j=1}^{k} x_{ij}\beta_j\right)\right)^2 - \left(g_1^{-1}\left(\sum_{j=1}^{k} x_{ij}\beta_j^0\right)\right)^2 - 2\sum_{i=1}^{n} g_1^{-1} \times \right.
$$

$$
\times \left. \left(\sum_{j=1}^{k} x_{ij}\beta_j^0\right) \left(g_1^{-1}\left(\sum_{j=1}^{k} x_{ij}\beta_j\right) - g_1^{-1}\left(\sum_{j=1}^{k} x_{ij}\beta_j^0\right)\right) \right] +
$$

$$
+ \frac{1}{n} \sum_{i=1}^{n} \left[y_i - g_1^{-1}\left(\sum_{j=1}^{k} x_{ij}\beta_j\right) \right]^2 \left[\frac{1}{g_2^{-1}\left(\sum_{j=1}^{l} z_{ij}\gamma_j\right)} - \right.
$$

$$
\left. - \frac{1}{g_2^{-1}\left(\sum_{j=1}^{l} z_{ij}\gamma_j^0\right)} \right] + \frac{1}{n} \sum_{i=1}^{n} \log \frac{g_2^{-1}\left(\sum_{j=1}^{l} z_{ij}\gamma_j\right)}{g_2^{-1}\left(\sum_{j=1}^{l} z_{ij}\gamma_j^0\right)} \right\} =
$$

$$
= \lim_{n\to\infty} \left\{ \frac{1}{n} \sum_{i=1}^{n} \frac{1}{g_2^{-1}\left(\sum_{j=1}^{l} z_{ij}\gamma_j\right)} \left[\left(g_1^{-1}\left(\sum_{j=1}^{k} x_{ij}\beta_j\right)\right)^2 + \right. \right.
$$

$$
+ \left(g_1^{-1}\left(\sum_{j=1}^{k} x_{ij}\beta_j^0\right)\right)^2 - 2g_1^{-1}\left(\sum_{j=1}^{k} x_{ij}\beta_j\right) g_1^{-1}\left(\sum_{j=1}^{k} x_{ij}\beta_j^0\right) \right] +
$$

$$
+ \; \frac{1}{n} \sum_{i=1}^{n} \left[y_i - g_1^{-1} \left(\sum_{j=1}^{k} x_{ij} \beta_j \right) \right]^2 \left[\frac{1}{g_2^{-1} \left(\sum_{j=1}^{l} z_{ij} \gamma_j \right)} - \right.
$$

$$
\left. - \; \frac{1}{g_2^{-1} \left(\sum_{j=1}^{l} z_{ij} \gamma_j^0 \right)} \right] + \frac{1}{n} \sum_{i=1}^{n} \log \frac{g_2^{-1} \left(\sum_{j=1}^{l} z_{ij} \gamma_j \right)}{g_2^{-1} \left(\sum_{j=1}^{l} z_{ij} \gamma_j^0 \right)} \Bigg\} =
$$

$$
= \lim_{n \to \infty} \left\{ \frac{1}{n} \sum_{i=1}^{n} \frac{1}{g_2^{-1} \left(\sum_{j=1}^{l} z_{ij} \gamma_j \right)} \left[g_1^{-1} \left(\sum_{j=1}^{k} x_{ij} \beta_j \right) - \right. \right.
$$

$$
\left. - \; g_1^{-1} \left(\sum_{j=1}^{k} x_{ij} \beta_j^0 \right) \right]^2 + \frac{1}{n} \sum_{i=1}^{n} \left[y_i - g_1^{-1} \left(\sum_{j=1}^{k} x_{ij} \beta_j \right) \right]^2 \times
$$

$$
\times \; \left[\frac{1}{g_2^{-1} \left(\sum_{j=1}^{l} z_{ij} \gamma_j \right)} - \frac{1}{g_2^{-1} \left(\sum_{j=1}^{l} z_{ij} \gamma_j^0 \right)} \right] +
$$

$$
+ \; \frac{1}{n} \sum_{i=1}^{n} \log \frac{g_2^{-1} \left(\sum_{j=1}^{l} z_{ij} \gamma_j \right)}{g_2^{-1} \left(\sum_{j=1}^{l} z_{ij} \gamma_j^0 \right)} \Bigg\}. \tag{3.6}
$$

Consider the first term on the right-hand side of (3.6). Since the function $g_1(t)$ is strictly monotone, then $g_1(t_1) \neq g_1(t_2)$ for $t_1 \neq t_2$. Since $\vec{\beta} \neq \vec{\beta}^0$, there exists a parallelepiped $A \subset B$ such that for all $\delta < \| \vec{\beta} - \vec{\beta}^0 \| / C_{\vec{x}}$ and $\vec{x} \in A$

$$\left| \sum_{j=1}^{k} x_j (\beta_j - \beta_j^0) \right| > \delta, \quad \vec{x} = (x_1, \dots, x_k),$$

hence

$$g_1^{-1} \left(\sum_{j=1}^{k} x_j \beta_j \right) \neq g_1^{-1} \left(\sum_{j=1}^{k} x_j \beta_j^0 \right), \quad \vec{x} \in A. \tag{3.7}$$

Taking into account (3.7), we get

$$\frac{1}{n} \sum_{i=1}^{n} \frac{1}{g_2^{-1} \left(\sum_{j=1}^{l} z_{ij} \gamma_j \right)} \left[g_1^{-1} \left(\sum_{j=1}^{k} x_{ij} \beta_j \right) - g_1^{-1} \left(\sum_{j=1}^{k} x_{ij} \beta_j^0 \right) \right]^2 \geq$$

$$\geq \frac{1}{n} \sum_{i=1}^{n} \frac{1}{g_2^{-1} \left(\sum_{j=1}^{l} z_{ij} \gamma_j \right)} \left[g_1^{-1} \left(\sum_{j=1}^{k} x_{ij} \beta_j \right) - g_1^{-1} \left(\sum_{j=1}^{k} x_{ij} \beta_j^0 \right) \right]^2 \times$$

$$\times \ \chi_{[\vec{x}_i \in A]} \geq \frac{1}{\alpha n} \sum_{i=1}^{n} \chi_{[\vec{x}_i \in A]} \inf_{\vec{x} \in A} \left[g_1^{-1} \left(\sum_{j=1}^{k} x_{ij} \beta_j \right) - \right.$$

$$\left. - \ g_1^{-1} \left(\sum_{j=1}^{k} x_{ij} \beta_j^0 \right) \right]^2 \geq \frac{1}{\alpha} \inf_{\vec{x} \in A} \left[g_1^{-1} \left(\sum_{j=1}^{k} x_j \beta_j \right) - \right.$$

$$\left. - \ g_1^{-1} \left(\sum_{j=1}^{k} x_j \beta_j^0 \right) \right]^2 \frac{1}{n} \sum_{i=1}^{n} \chi_{[\vec{x}_i \in A]}.$$

It is clear that

$$\inf_{\vec{x} \in A} \left[g_1^{-1} \left(\sum_{j=1}^{k} x_j \beta_j \right) - g_1^{-1} \left(\sum_{j=1}^{k} x_j \beta_j^0 \right) \right]^2 > 0.$$

We obtain from (3.1) that

$$\lim_{n \to \infty} n^{-1} \sum_{i=1}^{n} \chi_{[\vec{x}_i \in A]} > 0.$$

Hence

$$\lim_{n \to \infty} \frac{1}{n} \sum_{i=1}^{n} \frac{1}{g_2^{-1}\left(\sum_{j=1}^{l} z_{ij}\gamma_j\right)} \left[g_1^{-1}\left(\sum_{j=1}^{k} x_{ij}\beta_j\right) - g_1^{-1}\left(\sum_{j=1}^{k} x_{ij}\beta_j^0\right) \right]^2 > 0.$$

Consider the second and third terms on the right-hand side of (3.6). It is clear that

$$\alpha_n(\vec{\beta}^0, \vec{\gamma}) = \frac{1}{n} \sum_{i=1}^{n} \left[y_i - g_1^{-1}\left(\sum_{j=1}^{k} x_{ij}\beta_j\right) \right]^2 \times$$

$$\times \left[\frac{1}{g_2^{-1}\left(\sum_{j=1}^{l} z_{ij}\gamma_j\right)} - \frac{1}{g_2^{-1}\left(\sum_{j=1}^{l} z_{ij}\gamma_j^0\right)} \right] + \frac{1}{n} \sum_{i=1}^{n} \log \frac{g_2^{-1}\left(\sum_{j=1}^{l} z_{ij}\gamma_j\right)}{g_2^{-1}\left(\sum_{j=1}^{l} z_{ij}\gamma_j^0\right)} =$$

$$= -\frac{1}{n} \sum_{i=1}^{n} \log \frac{f\left(y_i, g_1^{-1}\left(\sum_{j=1}^{k} x_j\beta_j\right), g_2^{-1}\left(\sum_{j=1}^{l} z_{ij}\gamma_j\right)\right)}{f\left(y_i, g_1^{-1}\left(\sum_{j=1}^{k} x_j\beta_j\right), g_2^{-1}\left(\sum_{j=1}^{l} z_{ij}\gamma_j^0\right)\right)}, \tag{3.8}$$

where $f(x, \mu, \sigma^2)$ is the probability density of Gaussian distribution $\mathcal{N}(\mu, \sigma^2)$.

Since $E \log |x| < \log E |x|$, then we obtain

$$E \log \frac{f\left(y_i, g_1^{-1}\left(\sum_{j=1}^{k} x_j\beta_j\right), g_2^{-1}\left(\sum_{j=1}^{l} z_{ij}\gamma_j\right)\right)}{f\left(y_i, g_1^{-1}\left(\sum_{j=1}^{k} x_j\beta_j\right), g_2^{-1}\left(\sum_{j=1}^{l} z_{ij}\gamma_j^0\right)\right)} \leq$$

$$\leq \log E \, \frac{f\left(y_i, g_1^{-1}\left(\sum_{j=1}^{k} x_j\beta_j\right), g_2^{-1}\left(\sum_{j=1}^{l} z_{ij}\gamma_j\right)\right)}{f\left(y_i, g_1^{-1}\left(\sum_{j=1}^{k} x_j\beta_j\right), g_2^{-1}\left(\sum_{j=1}^{l} z_{ij}\gamma_j^0\right)\right)} = \log 1 = 0.$$

On the other hand

$$E \log \, \frac{f\left(y_i, g_1^{-1}\left(\sum_{j=1}^{k} x_j\beta_j\right), g_2^{-1}\left(\sum_{j=1}^{l} z_{ij}\gamma_j\right)\right)}{f\left(y_i, g_1^{-1}\left(\sum_{j=1}^{k} x_j\beta_j\right), g_2^{-1}\left(\sum_{j=1}^{l} z_{ij}\gamma_j^0\right)\right)} =$$

$$= \frac{g_2^{-1}\left(\sum_{j=1}^{l} z_{ij}\gamma_j^0\right) - g_2^{-1}\left(\sum_{j=1}^{l} z_{ij}\gamma_j\right)}{g_2^{-1}\left(\sum_{j=1}^{l} z_{ij}\gamma_j\right)} +$$

$$+ \log \, \frac{g_2^{-1}\left(\sum_{j=1}^{l} z_{ij}\gamma_j\right)}{g_2^{-1}\left(\sum_{j=1}^{l} z_{ij}\gamma_j^0\right)} = \varphi(j, \gamma_j, \gamma_j^0). \tag{3.9}$$

Thus $\varphi(j, \gamma_j, \gamma_j^0) > 0$ for $\gamma_j \neq \gamma_j^0$. The inequality

$$\lim_{n \to \infty} \alpha_n(\vec{\beta}^{\,0}, \vec{\gamma}) > 0$$

may be proved in the same manner as in the case of the first term of the right side in (3.6) by taking into account condition 2.

Therefore condition 1 of Lemma 13 is satisfied. Let us check condition 2 of this lemma. Let

$$X_{k,l}^{1,2}(\vec{\beta}, \vec{\gamma}, \delta) = \left\{ \| \vec{\beta}^{\,1} - \vec{\beta}^{\,2} \| + \| \vec{\gamma}^{\,1} - \vec{\gamma}^{\,2} \| \leq \delta \right\}.$$

Then

$$\sup_{X_{k,l}^{1,2}(\vec{\beta}, \vec{\gamma}, \delta)} \left| F_n(\vec{\beta}^{\,1}, \vec{\gamma}^{\,1}) - F_n(\vec{\beta}^{\,2}, \vec{\gamma}^{\,2}) \right| =$$

$$= \sup_{X^{1,2}_{k,l}\,(\vec{\beta},\vec{\gamma},\delta)} \left| \frac{1}{n}\sum_{i=1}^{n}\left\{ \frac{\left[y_i - g_1^{-1}\left(\sum_{j=1}^{k} x_{ij}\beta_j^1\right)\right]^2}{g_2^{-1}\left(\sum_{j=1}^{l} z_{ij}\gamma_j^1\right)} - \right.\right.$$

$$\left.\left. \frac{\left[y_i - g_1^{-1}\left(\sum_{j=1}^{k} x_{ij}\beta_j^2\right)\right]^2}{g_2^{-1}\left(\sum_{j=1}^{l} z_{ij}\gamma_j^2\right)} \right\} + \frac{1}{n}\sum_{i=1}^{n}\log \frac{g_2^{-1}\left(\sum_{j=1}^{l} z_{ij}\gamma_j^2\right)}{g_2^{-1}\left(\sum_{j=1}^{l} z_{ij}\gamma_j^1\right)} \right| =$$

$$= \sup_{X^{1,2}_{k,l}\,(\vec{\beta},\vec{\gamma},\delta)} \left| \frac{1}{n}\sum_{i=1}^{n}\left\{ \frac{\left[y_i - g_1^{-1}\left(\sum_{j=1}^{k} x_{ij}\beta_j^1\right)\right]^2}{g_2^{-1}\left(\sum_{j=1}^{l} z_{ij}\gamma_j^1\right)} - \right.\right.$$

$$- \frac{\left[y_i - g_1^{-1}\left(\sum_{j=1}^{k} x_{ij}\beta_j^2\right)\right]^2}{g_2^{-1}\left(\sum_{j=1}^{l} z_{ij}\gamma_j^1\right)} + \left[y_i - g_1^{-1}\left(\sum_{j=1}^{k} x_{ij}\beta_j^2\right)\right]^2 \times$$

$$\times \left[\frac{1}{g_2^{-1}\left(\sum_{j=1}^{l} z_{ij}\gamma_j^1\right)} - \frac{1}{g_2^{-1}\left(\sum_{j=1}^{l} z_{ij}\gamma_j^2\right)} \right] \left.\left.\right\}\right\} + \frac{1}{n}\sum_{i=1}^{n}\log \frac{g_2^{-1}\left(\sum_{j=1}^{l} z_{ij}\gamma_j^2\right)}{g_2^{-1}\left(\sum_{j=1}^{l} z_{ij}\gamma_j^1\right)} \right|.$$

$$(3.10)$$

Consider the first term in (3.10):

$$\varlimsup_{n\to\infty} \sup_{\|\vec{\beta}^1 - \vec{\beta}^2\|\le\delta} \times$$

$$\times \left| \frac{1}{n} \sum_{i=1}^{n} \frac{\left[y_i - g_1^{-1}\left(\sum_{j=1}^{k} x_{ij}\beta_j^1 \right) \right]^2 - \left[y_i - g_1^{-1}\left(\sum_{j=1}^{k} x_{ij}\beta_j^2 \right) \right]^2}{g_2^{-1}\left(\sum_{j=1}^{l} z_{ij}\gamma_j^1 \right)} \right| \le$$

$$\le \varlimsup_{n \to \infty} \sup_{\|\vec{\beta}^1 - \vec{\beta}^2\| \le \delta} \left| \frac{1}{n} \sum_{i=1}^{n} \frac{1}{g_2^{-1}\left(\sum_{j=1}^{l} z_{ij}\gamma_j^1 \right)} \left[g_1^{-1}\left(\sum_{j=1}^{k} x_{ij}\beta_j^1 \right) - \right. \right.$$

$$\left. \left. - \ g_1^{-1}\left(\sum_{j=1}^{k} x_{ij}\beta_j^2 \right) \right] \left[2y_i - g_1^{-1}\left(\sum_{j=1}^{k} x_{ij}\beta_j^1 \right) - g_1^{-1}\left(\sum_{j=1}^{k} x_{ij}\beta_j^2 \right) \right] \right| \le$$

$$\le \varlimsup_{n \to \infty} \frac{1}{n} \sum_{i=1}^{n} \sup_{\|\vec{\beta}^1 - \vec{\beta}^2\| \le \delta} \left\{ \frac{\left| g_1^{-1}\left(\sum_{j=1}^{k} x_{ij}\beta_j^1 \right) - g_1^{-1}\left(\sum_{j=1}^{k} x_{ij}\beta_j^2 \right) \right|}{\alpha} \times \right.$$

$$\times \ \left. \Big(|y_i| + C_1 + C_2 \Big) \right\} \le$$

$$\le \varlimsup_{n \to \infty} \frac{1}{n} \sum_{i=1}^{n} \sup_{\|\vec{\beta}^1 - \vec{\beta}^2\| \le \delta} \left\{ \frac{\left| g_1^{-1}\left(\sum_{j=1}^{k} x_{ij}\beta_j^1 \right) - g_1^{-1}\left(\sum_{j=1}^{k} x_{ij}\beta_j^2 \right) \right|}{\alpha} \times \right.$$

$$\times \ \left. \Big(E|y_i| + C_1 + C_2 \Big) \right\} \le$$

$$\le C \varlimsup_{n \to \infty} \frac{1}{n} \sum_{i=1}^{n} \sup_{\|\vec{\beta}^1 - \vec{\beta}^2\| \le \delta} \left| g_1^{-1}\left(\sum_{j=1}^{k} x_{ij}\beta_j^1 \right) - g_1^{-1}\left(\sum_{j=1}^{k} x_{ij}\beta_j^2 \right) \right| \le$$

$$\leq \; C \varlimsup_{n\to\infty} \frac{1}{n} \sum_{i=1}^{n} \sup_{\left|\sum_{j=1}^{k} x_{ij}(\vec{\beta}^1 - \vec{\beta}^2)\right| \leq C_{\vec{x}}\delta} \left| g_1^{-1}\left(\sum_{j=1}^{k} x_{ij}\beta_j^1\right) - \right.$$

$$\left. - \; g_1^{-1}\left(\sum_{j=1}^{k} x_{ij}\beta_j^2\right) \right| = C \varlimsup_{n\to\infty} \frac{1}{n} \sum_{i=1}^{n} \sup_{|t_1-t_2|\leq\varepsilon} \left| g_1^{-1}(t_1) - g_1^{-1}(t_2) \right|,$$

$$(3.11)$$

where $\varepsilon = C_{\vec{x}}\delta$.

It is obvious that the right-hand side of (3.11) converges to zero as $\delta \to 0$, since $g_1^{-1}(t)$ is continuous. Consider the second term on the right-hand side of (3.10):

$$\varlimsup_{n\to\infty} \sup_{\|\vec{\beta}^1 - \vec{\beta}^2\|\leq\delta} \left| \frac{1}{n} \sum_{i=1}^{n} \left[y_i - g_1^{-1}\left(\sum_{j=1}^{k} x_{ij}\beta_j^2\right) \right] \right| \times$$

$$\times \frac{\left| g_2^{-1}\left(\sum_{j=1}^{l} z_{ij}\gamma_j^2\right) - g_2^{-1}\left(\sum_{j=1}^{l} z_{ij}\gamma_j^1\right) \right|}{g_2^{-1}\left(\sum_{j=1}^{l} z_{ij}\gamma_j^1\right) g_2^{-1}\left(\sum_{j=1}^{l} z_{ij}\gamma_j^2\right)} \leq \varlimsup_{n\to\infty} \frac{1}{n} \sum_{i=1}^{n} \frac{\left| y_i - g_1^{-1}\left(\sum_{j=1}^{k} x_{ij}\beta_j^2\right) \right|}{\alpha^2} \times$$

$$\times \sup_{\|\vec{\gamma}^1 - \vec{\gamma}^2\|\leq\delta} \left| g_2^{-1}\left(\sum_{j=1}^{l} z_{ij}\gamma_j^2\right) - g_2^{-1}\left(\sum_{j=1}^{l} z_{ij}\gamma_j^1\right) \right| \leq$$

$$\leq \varlimsup_{n\to\infty} \frac{1}{n} \sum_{i=1}^{n} \sup_{\|\vec{\gamma}^1 - \vec{\gamma}^2\|\leq\delta} \left| g_2^{-1}\left(\sum_{j=1}^{l} z_{ij}\gamma_j^2\right) - g_2^{-1}\left(\sum_{j=1}^{l} z_{ij}\gamma_j^1\right) \right| =$$

$$= C \varlimsup_{n\to\infty} \frac{1}{n} \sum_{i=1}^{n} \sup_{|t_1-t_2|\leq\varepsilon} \left| g_2^{-1}(t_1) - g_2^{-1}(t_2) \right|, \qquad (3.12)$$

where $\varepsilon = C_z\delta$. The right-hand side of (3.12) converges to zero as $\delta \to 0$, since the function $g_2^{-1}(t)$ is continuous.

Now we consider the third term in (3.10):

$$\varlimsup_{n\to\infty}\frac{1}{n}\sum_{i=1}^{n}\sup_{\|\vec{\gamma}^1-\vec{\gamma}^2\|\le\delta}\left|\frac{1}{n}\sum_{i=1}^{n}\log\frac{g_2^{-1}\left(\sum_{j=1}^{l}z_{ij}\gamma_j^2\right)}{g_2^{-1}\left(\sum_{j=1}^{l}z_{ij}\gamma_j^1\right)}\right|\le$$

$$\varlimsup_{n\to\infty}\frac{1}{n}\sum_{i=1}^{n}\sup_{\|\vec{\gamma}^1-\vec{\gamma}^2\|\le\delta}\left|\log g_2^{-1}\left(\sum_{j=1}^{l}z_{ij}\gamma_j^2\right)-\right.$$

$$\left.-\log g_2^{-1}\left(\sum_{j=1}^{l}z_{ij}\gamma_j^1\right)\right|. \tag{3.13}$$

As above, the right-hand side of (3.12) converges to zero as $\delta \to 0$. The theorem is proved.

Now let us consider nonidentically distributed Gaussian random variables which satisfy a strong mixing condition.

Suppose that $\{\xi_i,\ i \in \mathcal{N}\}$ is Gaussian sequence of real random variables on the probabilistic space $(\Omega, \mathcal{G}, P)$, i. e., any finite collection from this sequence has joint Gaussian distribution. Denote

$$\mu_i = E\xi_i, \quad \sigma_i^2 = E(\xi_i - \mu_i)^2.$$

Assume that

$$g(\mu_i) = \vec{x}_i\,\vec{\beta}_0, \quad i \in \mathcal{N},$$

where $\vec{\beta}_0 \in \mathcal{R}^p$, $p \ge 1$, $\left\{\vec{x}_i,\ i \in \mathcal{N}\right\}$ is known sequence of vectors from $\mathcal{R}^p$, $\vec{x}_i\,\vec{\beta}_0$ is the scalar product of vectors $\vec{x}_i$ and $\vec{\beta}_0$, $g : \mathcal{R} \to \mathcal{R}$ is the continuous and strictly monotonic function. We need to estimate $\vec{\beta}_0$ by the observations

$$\left\{\xi_i = \xi_i(\omega), \quad i = \overline{1, n}\right\}.$$

Because the function g is continuous and strictly monotonic, it has the inverse function $g^{-1} : \mathcal{R} \to \mathcal{R}$, which is also continuous and strictly monotonic. We shall consider the least squares estimate, which is the solution to the problem

$$I_n(\vec{\beta}) = I_n(\vec{\beta}, \omega) = \frac{1}{n} \sum_{i=1}^{n} \frac{\left(\xi_i - g^{-1}(\vec{x}_i\,\vec{\beta})\right)^2}{\sigma_i^2} \to \min, \quad \vec{\beta} \in B, \qquad (3.14)$$

where B is a compact subset of $\mathcal{R}^p$.

As we can see the function $I_n(\vec{\beta})$ is also continuous in $\vec{\beta}$ for all n, ω. Then there exists at least one solution $\vec{\beta}_n = \vec{\beta}_n(\omega)$ to problem (3.14). Because of Theorem 6 the function $\vec{\beta}_n(\omega)$ can be chosen measurable for each n.

Lemma 14 *Suppose that* $\{\eta_i, i \in \mathcal{N}\}$ *is a sequence of real random variables with a finite second moment, defined on a probabilistic space* $(\Omega, \mathcal{G}, P)$. *Denote* $\zeta_i = \eta_i - E\eta_i$. *Assume that* $E(\eta_i)^2 \le K$, $i \in \mathcal{N}$, *and there exist the constant* c *and* $\varepsilon > 0$ *such that*

$$\sup_{i \in \mathcal{N}} \; \sup_{\substack{\xi_1 \in \mathcal{M}(\mathcal{F}_1^i) \\ \xi_2 \in \mathcal{M}(\mathcal{F}_{i+\tau}^\infty)}} \frac{|E(\xi_1 - E\xi_1)(\xi_2 - E\xi_2)|}{(E(\xi_1 - E\xi_1)^2 \, E(\xi_2 - E\xi_2)^2)^{1/2}} \le \frac{c}{\tau^{1+\varepsilon}}, \qquad (3.15)$$

where $\mathcal{F}_a^b = \sigma\{\eta_j, a \le j \le b\}$, $\mathcal{M}(\mathcal{F})$ *is the set of all random variables with finite second moments measurable with respect to* σ*-algebra* $\mathcal{F}$.

Then

$$P\left\{\alpha_n = \frac{1}{n} \sum_{i=1}^{n} \zeta_i \to 0, \quad n \to \infty\right\} = 1.$$

Proof. We obtain

$$E(\alpha_n)^2 = E\left(\frac{1}{n^2} \sum_{i,j=1}^{n} \zeta_i \zeta_j\right) = \frac{1}{n^2} \sum_{i,j=1}^{n} E\zeta_i \zeta_j =$$

$$= \frac{1}{n^2}\left(\sum_{i=1}^{n} E(\zeta_i)^2 + 2 \sum_{i=2}^{n} \sum_{j=1}^{i-1} E\zeta_i \zeta_j\right).$$

For $j < i$ we have

$$|E \zeta_i \zeta_j| \leq \frac{c}{(i-j)^{1+\varepsilon}} K.$$

Furthermore

$$\sum_{i=2}^{n} \sum_{j=1}^{i-1} \frac{1}{(i-j)^{1+\varepsilon}} = \sum_{i=2}^{n} \sum_{k=1}^{i-1} \frac{1}{k^{1+\varepsilon}} = \sum_{k=1}^{n-1} \sum_{i=k+1}^{n} \frac{1}{k^{1+\varepsilon}} =$$

$$= \sum_{k=1}^{n-1} \frac{n-k}{k^{1+\varepsilon}} \leq n \sum_{k=1}^{\infty} \frac{1}{k^{1+\varepsilon}} = nc_1.$$

Hence

$$E(\alpha_n)^2 \leq \frac{1}{n^2} \left(nK + 2cK \sum_{i=2}^{n} \sum_{j=1}^{i-1} \frac{1}{(i-j)^{1+\varepsilon}} \right) \leq$$

$$\leq \frac{1}{n^2} (nK + 2cKnc_1) = \frac{K(1+2cc_1)}{n} = \frac{c_2}{n}.$$

Denote $\gamma_m = \alpha_{m^2}$. Then $E(\gamma_m)^2 \leq c_2/(m^2)$.

Because of the Borel-Cantelli lemma

$$P\{\gamma_m \to 0, \quad m \to \infty\} = 1. \qquad (3.16)$$

For all n_0

$$\sup_{n \geq n_0} |\alpha_n| \leq \sup_{m:(m+1)^2 > n_0} \max_{m^2 \leq n < (m+1)^2} |\alpha_n| \leq$$

$$\leq \sup_{m:(m+1)^2 > n_0} \max_{m^2 \leq n < (m+1)^2} (|\alpha_n - \alpha_{m^2}| + |\alpha_{m^2}|) \leq$$

$$\leq \sup_{m:(m+1)^2 > n_0} \max_{m^2 \leq n < (m+1)^2} |\alpha_n - \alpha_{m^2}| + \sup_{m:(m+1)^2 > n_0} |\alpha_{m^2}|. \qquad (3.17)$$

Because of (3.16) the last term on the right-hand side of (3.17) almost surely converges to 0 as $n \to \infty$.

For $m^2 < n < (m+1)^2$

$$\alpha_n - \alpha_{m^2} = \frac{1}{n}\sum_{i=1}^{n}\zeta_i - \frac{1}{m^2}\sum_{i=1}^{m^2}\zeta_i =$$

$$= \frac{1}{n}\sum_{i=m^2+1}^{n}\zeta_i + \left(\frac{1}{n} - \frac{1}{m^2}\right)\sum_{i=1}^{m^2}\zeta_i. \qquad (3.18)$$

By virtue of the Cauchy-Buniakowski inequality,

$$E\left(\max_{m^2<n<(m+1)^2}\left|\frac{1}{n}\sum_{i=m^2+1}^{n}\zeta_i\right|\right)^2 \leq$$

$$\leq E \max_{m^2<n<(m+1)^2} \frac{1}{n^2}\sum_{i=m^2+1}^{n}(\zeta_i)^2 \sum_{i=m^2+1}^{n}1 \leq$$

$$\leq \frac{2m}{m^4}\sum_{i=m^2+1}^{(m+1)^2-1}E\,(\zeta_i)^2 \leq \frac{K(2m)^2}{m^4} = \frac{4K}{m^2}. \qquad (3.19)$$

Then

$$\max_{m^2\leq n<(m+1)^2}\left|\left(\frac{1}{n} - \frac{1}{m^2}\right)\sum_{i=1}^{m^2}\zeta_i\right| \leq \frac{1}{m^2}\left|\sum_{i=1}^{m^2}\zeta_i\right| =$$

$$= |\alpha_{m^2}| \overset{a.s.}{\to} 0, \quad n \to \infty. \qquad (3.20)$$

Because of (3.18) - (3.20)

$$P\left\{\max_{m^2\leq n<(m+1)^2}|\alpha_n - \alpha_{m^2}| \to 0, \quad m \to \infty\right\} = 1.$$

Then (3.17) implies the validity of the lemma.

Now we shall consider our problem.

Theorem 25 *Suppose that the following conditions are valid:*

1) g^{-1} is Lipschitz function;

2) $\vec{x} \leq C, i \in \mathcal{N};$

3) $0 < a \leq \sigma_i^2 \leq b, i \in \mathcal{N};$

4) the sequence $\{\xi_i, i \in \mathcal{N}\}$ satisfies the strong mixing condition with the coefficient $\alpha(k) = O(k^{-1-\varepsilon}), \varepsilon > 0;$

5) there exists r such that for all open balls $A \subset B_r(\vec{0})$

$$\lim_{n \to \infty} \frac{1}{n} \sum_{i=1}^{n} \chi(\vec{x}_i \in A) = \nu(A) > 0,$$

where

$$B_r(\vec{0}) = \left\{ \vec{x} \in \mathcal{R}^p : \| \vec{x} \| < r \right\},$$

$$\chi(\vec{x}_i \in A) = \begin{cases} 1, & \vec{x}_i \in A \\ 0, & \vec{x}_i \bar{\in} A \end{cases}.$$

Denote $\vec{\beta}_n = \vec{\beta}_n(\omega)$ some measurable solution to problem (3.14). Then

$$P\left\{ \vec{\beta}_n \to \vec{\beta}_0, \ n \to \infty \right\} = 1.$$

Proof. For all n the function $I_n(\vec{\beta}) = I_n(\vec{\beta}, \omega)$ is a normal integrant on $B \times \Omega$. We will check conditions 1 and 2 of Lemma 13 for the functions I_n.

For each $\vec{\beta}, \vec{\beta}_1 \in B$ denote

$$m_i = g^{-1}(\vec{x}_i \vec{\beta}), \quad q_i = g^{-1}(\vec{x}_i \vec{\beta}_1).$$

Then

$$I_n(\vec{\beta}) - I_n(\vec{\beta}_1) = \frac{1}{n} \sum_{i=1}^{n} \frac{(\xi_i - m_i)^2}{\sigma_i^2} - \frac{1}{n} \sum_{i=1}^{n} \frac{(\xi_i - q_i)^2}{\sigma_i^2} =$$

$$= \frac{1}{n} \sum_{i=1}^{n} \frac{(q_i - m_i)^2 + 2(q_i - m_i)(\xi_i - q_i)}{\sigma_i^2}. \quad (3.21)$$

Denote $\tilde{I}_n(\vec{\beta}) = I_n(\vec{\beta}) - I_n(\vec{\beta}_0)$. We obtain

$$E\,\tilde{I}_n(\vec{\beta}) = \frac{1}{n}\sum_{i=1}^{n}\frac{(\mu_i - m_i)^2}{\sigma_i^2},$$

$$\tilde{I}_n(\vec{\beta}) - E\,\tilde{I}_n(\vec{\beta}) = \frac{2}{n}\sum_{i=1}^{n}\frac{(\mu_i - m_i)\,(\xi_i - \mu_i)}{\sigma_i^2}.$$

Denote $\eta_i = (\mu_i - m_i)\,(\xi_i - \mu_i)/\sigma_i^2$. Let us check conditions of Lemma 14 for the sequence η_i. Then

$$E\,\eta_i = 0, \quad E\,(\eta_i)^2 = (\mu_i - m_i)\,(\xi_i - \mu_i)/\sigma_i^2.$$

Because of compactness of B

$$\|\,\vec{\beta}\,\| \le C_0, \quad \vec{\beta}\in B.$$

Then for any i, $\vec{\beta}$

$$|\,\vec{x}_i\,\vec{\beta}\,| \le \|\,\vec{x}_i\,\|\,\|\,\vec{\beta}\,\| \le CC_0,$$

$$\left|\,g^{-1}\,(\vec{x}_i\,\vec{\beta})\,\right| \le K. \tag{3.22}$$

Hence $E\,(\eta_i)^2 \le 4K^2/a$.

In accordance with [116], condition 4 of the theorem implies the validity of relation (3.15) for the sequence ξ_i. Because η_i is the measurable function with respect to ξ_i, for all a_1, a_2

$$\sigma\,\{\,\eta_j,\, a_1 \le j \le a_2\,\} \subset \sigma\,\{\,\xi_j,\, a_1 \le j \le a_2\,\}.$$

Then relation (3.15) is fulfilled for the sequence η_i, and all conditions of Lemma 14 are satisfied. We have

$$P\left\{\,\tilde{I}_n(\vec{\beta}) - E\,\tilde{I}_n(\vec{\beta}) \to 0, \quad n \to \infty\right\} = 1.$$

If $\vec{\beta}\neq\vec{\beta}_0$ then $(\vec{\beta} - \vec{\beta}_0)\,(\vec{\beta} - \vec{\beta}_0) > 0$ and for some λ

$$\lambda(\vec{\beta} - \vec{\beta}_0) \in B_r(\vec{0}), \quad \lambda(\vec{\beta} - \vec{\beta}_0)\,(\vec{\beta} - \vec{\beta}_0) > 0.$$

Denote

$$\vec{z} = \lambda(\vec{\beta} - \vec{\beta}_0), \quad \Phi(\vec{u}) = \left(g^{-1}(\vec{u}\,\vec{\beta}) - g^{-1}(\vec{u}\,\vec{\beta}_0) \right)^2.$$

The relation $\vec{z}\,(\vec{\beta} - \vec{\beta}_0) > 0$ implies that $\vec{z}\,\vec{\beta} \neq \vec{z}\,\vec{\beta}_0$. Because of the strict monotonicity of g^{-1},

$$g^{-1}(\vec{z}\,\vec{\beta}) \neq g^{-1}(\vec{z}\,\vec{\beta}_0).$$

Hence $\Phi(\vec{z}) = \gamma > 0$. Since Φ is continuous, there exists $\delta > 0$ such that for

$$\vec{u} \in A = B_\delta(\vec{z}) = \left\{ \vec{u} : \| \vec{u} - \vec{z} \| < \delta \right\}$$

we have $\Phi(\vec{u}) > \gamma/2$. Because $\vec{z} \in B_r(\vec{0})$ we obtain $A \subset B_r(\vec{0})$. Then

$$E\,\tilde{I}_n(\vec{\beta}) = \frac{1}{n} \sum_{i=1}^{n} \frac{\Phi(\vec{x}_i)}{\sigma_i^2} \geq \frac{\gamma}{2b} \frac{1}{n} \sum_{i=1}^{n} \chi(\vec{x}_i \in A),$$

$$\lim_{n \to \infty} E\,\tilde{I}_n(\vec{\beta}) = \frac{\gamma}{2b} \nu(A) > 0,$$

$$P\left\{ \lim_{n \to \infty} \tilde{I}_n(\vec{\beta}) \geq \lim_{n \to \infty} \left(\tilde{I}_n(\vec{\beta}) - E\,\tilde{I}_n(\vec{\beta}) \right) + \lim_{n \to \infty} E\,\tilde{I}_n(\vec{\beta}) > 0 \right\} = 1.$$

Then condition 1 of Lemma 13 is satisfied. Fix $\vec{\beta}_1 \in B$. Let

$$\eta_i' = \sup_{\|\vec{\beta} - \vec{\beta}_1\| \leq \delta} \left| (q_i - m_i)^2 + 2(q_i - m_i)(\xi_i - q_i) \right|,$$

where $q_i = g^{-1}(\vec{x}_i\,\vec{\beta}_1)$, $m_i = g^{-1}(\vec{x}_i\,\vec{\beta})$. From (3.21) we have

$$\Delta_n(\delta) = \sup_{\|\vec{\beta} - \vec{\beta}_1\| \leq \delta} \left| I_n(\vec{\beta}) - I_n(\vec{\beta}_1) \right| \leq \frac{1}{an} \sum_{i=1}^{n} \eta_i'.$$

It is evident that η_i is the measurable function with respect to ξ_i. Let us try to apply Lemma 14 to the sequence η_i. The relationship (3.15) is fulfilled. Then

$$\left| (q_i - m_i)^2 + 2(q_i - m_i)(\xi_i - q_i) \right| \leq 2LC \left(|\xi_i| + K \right) \| \vec{\beta} - \vec{\beta}_1 \|,$$

where α is Lipschitz constant for g^{-1} and K is the constant from (3.22). Hence

$$\eta_i' \le 2C\delta\left(|\xi_i| + K\right), \qquad E\,\eta_i' \le 2LC\delta\left(E\,|\xi_i - \mu_i| + 2K\right) \le c_2\,\delta,$$

$$E\,(\eta_i')^2 \le c_3, \qquad\qquad E\,(\eta_i' - E\,\eta_i')^2 \le c_4.$$

Then all conditions of Lemma 14 are satisfied. Applying Lemma 14 we have

$$P\left\{\frac{1}{n}\sum_{i=1}^n (\eta_i' - E\,\eta_i') \to 0, \quad n \to \infty\right\} = 1.$$

Consequently,

$$\varliminf_{n\to\infty} \Delta_n(\delta) \le \frac{1}{a}\varlimsup_{n\to\infty}\left(\frac{1}{n}\sum_{i=1}^n (\eta_i' - E\,\eta_i') + \frac{1}{n}\sum_{i=1}^n E\,\eta_i'\right) \le$$

$$\le \frac{1}{a}\varlimsup_{n\to\infty}\frac{1}{n}\sum_{i=1}^n E\,\eta_i' \le c_2\delta/a \to 0, \quad \delta \to 0$$

with probability 1. We see, that all conditions of Lemma 13 are fulfilled. This implies the validity of the theorem.

3.2 ESTIMATES OF THE PARAMETERS FOR GAUSSIAN RANDOM FIELD WITH A CONTINUOUS ARGUMENT

Let us now consider the properties of an estimator of the parameters of the linear regression on Gaussian homogeneous random field.

Let $\left\{x(\vec{t}), \vec{y}\,(\vec{t}), \vec{t} \in \mathcal{R}^m\right\}$, $m \ge 2$ be Gaussian homogeneous random field given on a complete probabilistic space $(\Omega, \mathcal{G}, P)$; $x(\vec{t}) \in \mathcal{R}$, $\vec{y}\,(\vec{t}) \in \mathcal{R}^l$, $\vec{t} \in \mathcal{R}^m$; $l \ge 1$.

The following assumptions will be used later.

1. For any $\vec{t} \in \mathcal{R}^m$

$$E\left\{x(\vec{t})\,/\,F\right\} = \left(\vec{y}\,(\vec{t})\right)' \vec{\theta}$$

with probability 1, where $F = \sigma \left\{ \vec{y}(\vec{\tau}), \ \vec{\tau} \in \mathcal{R}^m \right\}$; $\vec{\theta} \in \mathcal{R}^l$ is some fixed vector.

2. The traectories of the field $\left(x(\vec{t}), \ \vec{y}(\vec{t}) \right)$ are continuous with probability 1.

Denote

- $\vec{y}(\vec{t}) = \left(\vec{y}_j(\vec{t}) \right)_{j=1}^l$, $\quad \xi(\vec{t}) = \vec{x}(\vec{t}) - \left(\vec{y}(\vec{t}) \right)' \vec{\theta}, \quad \vec{t} \in \mathcal{R}^m$;

- $a_j = E\, y_j(\vec{0}), \quad j = \overline{1, l}, \quad \vec{a} = (a_j)_{j=1}^l$;

- $y_j^0(\vec{t}) = y_j(\vec{t}) - a_j, \quad \vec{t} \in \mathcal{R}^m, \quad j = \overline{1, l}$;

- $r_{jk}(\vec{t}) = E\, y_j^0(\vec{t})\, y_k^0(\vec{0}), \quad j, k = \overline{1, l}$,

- $r(\vec{t}) = E\, \xi(\vec{t})\, \xi(\vec{0}), \quad \vec{t} \in \mathcal{R}^m$.

It is worth to note that assumption 2 and properties of Gaussian homogeneous fields imply that functions $r_{jk}(\vec{t})$, $j, k = \overline{1, l}$ and $r(\vec{t})$ are continuous on $\mathcal{R}^m$.

3. The correlation functions satisfy the following conditions:

a) $c_{jk} = \displaystyle\int\limits_{\mathcal{R}^m} \left| r_{jk}(\vec{t}) \right| d\,\vec{t} < \infty; \ j, k = \overline{1, l}$;

b) for any set of indexes $I \subset \{ \overline{1, m} \}$, $I \neq \{ \overline{1, m} \}$,

$$c_{jk}(I) = \int\limits_{\mathcal{R}^{N(I)}} \left| r_{jk} \left(\vec{\varphi}(\vec{\tau}, I) \right) \right| d\,\vec{\tau} < \infty; \quad j, k = \overline{1, l},$$

where $N(I)$ is the number of elements in I; $\vec{\varphi}(\vec{\tau}, I)$ is the vector from $\mathcal{R}^m$, which has coordinates with indexes that do not belong to I, equal to zero, and coordinates with indexes from I – equal to corresponding coordinates of $\vec{\tau}$;

c) $c = \displaystyle\int\limits_{\mathcal{R}^m} \left| r(\vec{t}) \right| d\,\vec{t} < \infty$;

d) for each $I \subset \{ \overline{1, m} \}$, $I \neq \{ \overline{1, m} \}$,

$$c(I) = \int\limits_{\mathcal{R}^{N(I)}} \left| r \left(\vec{\varphi}(\vec{\tau}, I) \right) \right| d\,\vec{\tau} < \infty,$$

where designations were done in the same manner as in condition b).

4. The matrix $\left(r_{jk}(\vec{0})\right)_{j,k=1}^{l}$ is nonsingular.

Denote

$$G(\vec{T}) = \Pi[\,\vec{0},\vec{T}\,], \quad \vec{T} \in \mathcal{R}_+^m.$$

Then for any $\vec{T} = (T_i)_{i=1}^m \in \mathcal{R}^m$ the inequality $\vec{T} > 0$ will mean that $T_i > 0$, $i = \overline{1, m}$.

Let assumptions 1,2 be satisfied. We need to estimate the vector $\vec{\theta}$ by observations

$$\left\{\left(x(\vec{t}),\, y(\vec{t})\right),\; \vec{t} \in G(\vec{T})\right\}, \quad \vec{T} > 0.$$

Consider least squares estimate which is a point of minimum of the function

$$F_{\vec{T}}(\vec{u}) \;=\; F_{\vec{T}}(\vec{u},\omega) =$$

$$= \frac{1}{\prod\limits_{i=1}^{m} T_i} \int\limits_{G(\vec{T})} \left(x(\vec{t}) - \left(y(\vec{t})\right)'\vec{u}\right)^2 d\vec{t}, \quad \vec{u} \in \mathcal{R}^l, \quad (3.23)$$

where $\vec{T} = (T_i)_{i=1}^m$.

Using Lebesgue theorem of limit transition under an integral sign, it is easy to show that for all $\vec{T}$ the function (3.23) is twice differentiable in every point of $\mathcal{R}^l$ with probability 1. If the gradient of this function is equal to $\vec{0}$, then we have the equation

$$Q(\vec{T})\,\vec{u} = \frac{1}{\prod\limits_{i=1}^{m} T_i} \int\limits_{G(\vec{T})} x(\vec{t})\,\vec{y}\,(\vec{t})\,d\vec{t}, \quad (3.24)$$

where

$$Q(\vec{T}) = \frac{1}{\prod\limits_{i=1}^{m} T_i} \int\limits_{G(\vec{T})} \vec{y}\,(\vec{t}) \left(\vec{y}\,(\vec{t})\right)' d\vec{t}\,.$$

Then the matrix of partial derivatives of the second order of function (3.23) is equal to $2Q(\vec{T})$ in any point.

Let us investigate the properties of $Q(\vec{T})$.

Lemma 15 *If assumptions* 2, 3*a*) *hold then*

$$E \parallel Q(\vec{T}) - Q \parallel^2 \to 0, \quad \prod_{i=1}^{m} T_i \to \infty,$$

where

$$Q = E \left\{ \vec{y}\,(\vec{0}) \left(\vec{y}\,(\vec{0}) \right)' \right\} = \left(r_{jk}(\vec{0}) + a_j\,a_k \right)'_{j,k=1}.$$

If also 3*b*) *holds then*

$$P \left\{ Q(\vec{T}) \to Q, \quad \vec{T} \to \infty \right\} = 1.$$

Proof. Fix arbitrary j, k. Denote $q_{jk}(\vec{T})$ and q_{jk} elements with indexes j, k of matrices $Q(\vec{T})$ and Q respectively.

Fix $\vec{T} > 0$, then we have

$$\Delta_{jk}(\vec{T}) = E \left(q_{jk}(\vec{T}) - q_{jk} \right)^2 = E \left(\frac{1}{\prod\limits_{i=1}^{m} T_i} \int\limits_{G(\vec{T})} \left(y_j(\vec{t})\,y_k(\vec{t}) - q_{jk} \right) d\,\vec{t} \right)^2 =$$

$$= \frac{1}{\prod\limits_{i=1}^{m} (T_i)^2} E \int\limits_{\left(G(\vec{T}) \right)^2} \left(\left(y_j(\vec{t})\,y_k(\vec{t}) - q_{jk} \right) \left(y_j(\vec{s})\,y_k(\vec{s}) - q_{jk} \right) \right) d\,\vec{t}\ d\,\vec{s}\,.$$

Because Gaussian homogeneous field has restricted moments of any order, then using Fubini theorem and formulas for the moments of Gaussian random function [32]; *pp.*29, 30 we obtain

$$\Delta_{jk}(\vec{T}) = \frac{1}{\prod\limits_{i=1}^{m} (T_i)^2} \int\limits_{\left(G(\vec{T}) \right)^2} \alpha_{jk}(\vec{t} - \vec{s})\,d\,\vec{t}\ d\,\vec{s},$$

where

$$\alpha_{jk}(\vec{t}) \;=\; r_{jj}(\vec{t})\,r_{kk}(\vec{t}) + r_{jk}(\vec{t})\,r_{kj}(\vec{t}) + (\alpha_j)^2\,r_{kk}(\vec{t}) +$$

$$+\; \alpha_j\,\alpha_k\left(r_{jk}(\vec{t}) + r_{kj}(\vec{t})\right) + (\alpha_k)^2\,r_{jj}(\vec{t}), \quad \vec{t}\in\mathcal{R}^m. \quad (3.25)$$

After changing variables and using 3a) we have

$$\Delta_{jk}(\vec{T}) = \frac{1}{\prod\limits_{i=1}^{m}(T_i)^2} \int\limits_{G(\vec{T})} \left(\int\limits_{\Pi\,[\,-\vec{s},\vec{T}-\vec{s}\,]} \alpha_{jk}(\vec{t})\,d\,\vec{t} \right) d\,\vec{s} \le \frac{\alpha_{jk}}{\prod\limits_{i=1}^{m} T_i}, \quad (3.26)$$

where

$$\alpha_{jk} = r_{jj}(\vec{0})\,c_{kk} + \left(r_{jj}(\vec{0})\,r_{kk}(\vec{0})\right)^{1/2} c_{kj} + (a_j)^2\,c_{kk} + 2a_j\,a_k\,c_{jk} + (a_k)^2 c_{jj}.$$

Note that $c_{jk} = c_{kj}$ because $r_{jk}(\vec{t}) = r_{kj}(-\vec{t})$, $\vec{t}\in\mathcal{R}^m$.

Then the first part of the lemma is true.

For any vector $\vec{n} = (n_i)_{i=1}^{m} \in \mathcal{N}^m$ denote $\vec{n}^2 = ((n_i)^2)_{i=1}^{m}$. Because of (3.26)

$$E\left(q_{jk}(\vec{n}^2) - q_{jk}\right)^2 \le \frac{\alpha_{jk}}{\prod\limits_{i=1}^{m}(n_i)^2}, \quad \vec{n} = (n_i)_{i=1}^{m} \in \mathcal{N}^m.$$

Hence by virtue of Borel-Cantelli lemma

$$P\left\{q_{jk}(\vec{n}^2) \to q_{jk}, \quad \vec{n}\to\infty\right\} = 1.$$

For all $T_0 \ge 1$ we have

$$\sup_{\vec{T}=(T_i)_{i=1}^{m} \in\mathcal{R}^m\,:\,T_i>T_0,\,i=\overline{1,m}} |q_{jk}(\vec{T}) - q_{jk}| \le$$

$$\le \sup_{\vec{n}=(n_i)_{i=1}^{m} \in\mathcal{N}^m\,:\,(n_i+1)^2>T_0,\,i=\overline{1,m}} \;\; \sup_{\vec{T}\in G_0(\vec{n})} |q_{jk}(\vec{T}) - q_{jk}(\vec{n}^2)| +$$

$$+ \sup_{\vec{n}=(n_i)_{i=1}^{m} \in\mathcal{N}^m\,:\,(n_i+1)^2>T_0,\,i=\overline{1,m}} |q_{jk}(\vec{n}^2) - q_{jk}| \quad (3.27)$$

with probability 1, where

$$G_0(\vec{n}) = \left\{ \vec{T} = (T_i)_{i=1}^m \in \mathcal{R}^m : (n_i)^2 \le T_i < (n_i+1)^2, \, i = \overline{1,m} \right\},$$

$$\vec{n} = (n_i)_{i=1}^m \in \mathcal{N}^m.$$

Denote

$$\overline{I} = \left\{ \overline{1,m} \right\} \setminus I, \quad G_1(I, \vec{n}, \vec{T}) = \left\{ \vec{t} = (t_i)_{i=1}^m \in \mathcal{R}^m : 0 \le t_i \le (n_i)^2, \right.$$

$$i \in I; \quad (n_i)^2 \le t_i \le T_i, \, i \in \overline{I} \right\}, \quad I \subset \{i = \overline{1,m}\}, \quad \vec{n} = (n_i)_{i=1}^m \in \mathcal{N}^m,$$

$$\vec{T} = (T_i)_{i=1}^m \in G_0(\vec{n}).$$

For all $\vec{n} \in \mathcal{N}^m$, $\vec{T} \in G_0(\vec{n})$

$$\left| q_{jk}(\vec{T}) - q_{jk}(\vec{n}^2) \right| \le \frac{1}{\prod\limits_{i=1}^m T_i} \left| \int\limits_{G(\vec{T})} y_j(\vec{t}) \, y_k(\vec{t}) d\,\vec{t} \, - \right.$$

$$\left. - \int\limits_{G(\vec{n}^2)} y_j(\vec{t}) \, y_k(\vec{t}) d\,\vec{t} \right| + \left| \left(\frac{1}{\prod\limits_{i=1}^m T_i} - \frac{1}{\prod\limits_{i=1}^m (n_i)^2} \right) \int\limits_{G(\vec{n}^2)} y_j(\vec{t}) \, y_k(\vec{t}) d\,\vec{t} \right| \le$$

$$\le \frac{1}{\prod\limits_{i=1}^m (n_i)^2} \left| \int\limits_{G(\vec{T}) \setminus G(\vec{n}^2)} \left(y_j(\vec{t}) \, y_k(\vec{t}) - q_{jk} \right) d\,\vec{t} \right| +$$

$$+ \frac{|q_{jk}|}{\prod\limits_{i=1}^m (n_i)^2} \left(\int\limits_{G(\vec{T})} d\,\vec{t} \int\limits_{G(\vec{n}^2)} d\,\vec{t} \right) + \left| \left(\frac{1}{\prod\limits_{i=1}^m (T_i/(n_i)^2)} - 1 \right) q_{jk}(\vec{n}^2) \right| \le$$

$$\leq \frac{1}{\prod\limits_{i=1}^{m}(n_i)^2} \sum_{\substack{I\subset\{i=\overline{1,m}\} \\ I\neq\{i=\overline{1,m}\}}} \left| \int\limits_{G_1(I,\vec{n},\vec{T})} \left(y_j(\vec{t})\,y_k(\vec{t})-q_{jk}\right)d\,\vec{t} \right| +$$

$$+\frac{1}{\prod\limits_{i=1}^{m}(n_i)^2} \left| \int\limits_{\Pi[\vec{n}^2,\vec{T}]} \left(y_j(\vec{t})\,y_k(\vec{t})-q_{jk}\right)d\,\vec{t} \right| +$$

$$+|q_{jk}|\left(\prod_{i=1}^{m}(1+1/n_i)^2-1\right)+\left(1-\frac{1}{\prod\limits_{i=1}^{m}(1+1/n_i)^2}\right)\left|q_{jk}(\vec{n}^2)\right| \qquad (3.28)$$

with probability 1.

Denote

$$G_2(I,\vec{a},\vec{b})=\left\{\vec{\tau}=(\tau_i)_{i\in I}\in\mathcal{R}^{N(I)}:a_i\leq\tau_i\leq b_i,\,i\in I\right\};$$

$$I\subset\{\overline{1,m}\},\quad \vec{a}=(a_i)_{i=1}^{m}\quad \vec{b}=(b_i)_{i=1}^{m}\in\mathcal{R}^m.$$

Fix a set $I\subset\{\overline{1,m}\}$, $I\neq\{\overline{1,m}\}$, and $\vec{n}\in\mathcal{N}^m$. After applying Fubini theorem and Cauchy-Buniakowski inequality we obtain

$$\Delta_{jk}^1(\vec{n},I) = E\left(\sup_{\vec{T}\in G_0(\vec{n})}\frac{1}{\prod\limits_{i=1}^{m}(n_i)^2}\left|\int\limits_{G_1(I,\vec{n},\vec{T})}\left(y_j(\vec{t})\,y_k(\vec{t})-q_{jk}\right)d\,\vec{t}\right|\right)^2 \leq$$

$$\leq \frac{1}{\prod\limits_{i=1}^{m}(n_i)^4}E\sup_{\vec{T}\in G_0(\vec{n})}\left(\prod_{i\in\overline{I}}(T_i-(n_i)^2)\int\limits_{G_2(I,\vec{n}^2,\vec{T})}\times\right.$$

$$\times\left.\left(\left(y_j\left(\vec{\psi}(\vec{\tau},\vec{\nu})\right)y_k\left(\vec{\psi}(\vec{\tau},\vec{\nu})\right)-q_{jk}\right)d\,\vec{\tau}\right)^2d\,\vec{\nu}\right),$$

where $\vec{\psi}\,(\vec{\tau},\vec{\nu}) = \left(\psi_i(\vec{\tau},\vec{\nu})\right)_{i=1}^{m}$;

$$\psi_i(\vec{\tau},\vec{\nu}) = \begin{cases} \tau_i, & i \in I \\ \nu_i, & i \in \bar{I}, \end{cases} \qquad i = \overline{1,m},$$

$$\vec{\tau} = (\tau_i)_{i \in I} \in \mathcal{R}^{N(I)}, \qquad \vec{\nu} = (\nu_i)_{i \in \bar{I}} \in \mathcal{R}^{N(\bar{I})}.$$

Then we have

$$\Delta_{jk}^{1}(\vec{n},I) \leq \frac{\prod\limits_{i \in \bar{I}}(2n_i+1)}{\prod\limits_{i=1}^{m}(n_i)^4} E \int\limits_{G_2\left(\bar{I},\vec{n}^2,(\vec{n}+\vec{1})^2\right)} \times$$

$$\times \left(\int\limits_{\left(G_2(I,\vec{0},\vec{n}^2)\right)^2} \left(y_j\left(\vec{\psi}\,(\vec{\tau},\vec{\nu})\right) y_k\left(\vec{\psi}\,(\vec{\tau},\vec{\nu})\right) - q_{jk}\right) \times \right.$$

$$\left. \times \left(y_j\left(\vec{\psi}\,(\vec{\lambda},\vec{\nu})\right) y_k\left(\vec{\psi}\,(\vec{\lambda},\vec{\nu})\right) - q_{jk}\right) d\vec{\tau}\,d\vec{\lambda}\right) d\vec{\nu} =$$

$$= \frac{\prod\limits_{i \in \bar{I}}(2n_i+1)^2}{\prod\limits_{i=1}^{m}(n_i)^4} \int\limits_{\left(G_2(I,\vec{0},\vec{n}^2)\right)^2} \alpha_{jk}\left(\vec{\varphi}\,(\vec{\tau}-\vec{\lambda},I)\right) d\vec{\tau}\,d\vec{\lambda} =$$

$$= \frac{\prod\limits_{i \in \bar{I}}(2+1/n_i)^2}{\prod\limits_{i \in \bar{I}}(n_i)^2 \prod\limits_{i \in I}(n_i)^4} \int\limits_{G_2(I,\vec{0},\vec{n}^2)} \times$$

$$\times \left(\int\limits_{G_2\left(I,-\vec{\varphi}(\vec{\lambda},I),\vec{n}^2-\vec{\varphi}(\vec{\lambda},I)\right)} \alpha_{jk}\left(\vec{\varphi}\,(\vec{\tau},I)\right) d\vec{\tau}\right) d\vec{\lambda},$$

where $\vec{\varphi}\,(\vec{\tau}, I)$ is denoted in assumption 3b), and the function $\alpha_{jk}(\vec{t})$, $\vec{t} \in \mathcal{R}^m$ is defined by (3.25).

Assumption 3b) implies

$$\Delta^1_{jk}(\vec{n}, I) \leq \frac{\prod\limits_{i \in \overline{I}}(2 + 1/n_i)^2\, \alpha^0_{jk}(I)}{\prod\limits_{i=1}^{m}(n_i)^2} \leq \frac{\alpha_1}{\prod\limits_{i=1}^{m}(n_i)^2}, \qquad (3.29)$$

where $\alpha^0_{jk}(I) = r_{jj}(\vec{0})\, c_{kk}(I) + \left(r_{jj}(\vec{0})\, r_{kk}(\vec{0})\right)^{1/2} c_{kj}(I) + (a_j)^2\, c_{kk}(I) + 2a_j\, a_k\, c_{jk}(I) +$ $(a_k)^2\, c_{jj}(I)$; α_1 is a constant.

For any $\vec{n}$

$$E\left(\sup_{\vec{T} \in G_0(\vec{n})} \frac{1}{\prod\limits_{i=1}^{m}(n_i)^2} \left| \int\limits_{\Pi[\vec{n}^2, \vec{T}]} \left(y_j(\vec{t})\, y_k(\vec{t}) - q_{jk}\right) d\,\vec{t} \right| \right)^2 \leq$$

$$\leq \frac{1}{\prod\limits_{i=1}^{m}(n_i)^4} E \sup_{\vec{T} \in G_0(\vec{n})} \left(\prod\limits_{i=1}^{m}(T_i - (n_i)^2) \int\limits_{\Pi[\vec{n}^2, \vec{T}]} \left(y_j(\vec{t})\, y_k(\vec{t}) -\right.\right.$$

$$\left.\left. - q_{jk}\right)^2 d\,\vec{t} \right) = \alpha_{jk}(\vec{0}) \prod\limits_{i=1}^{m} \frac{(2n_i + 1)^2}{(n_i)^4} \leq \frac{\alpha_2}{\prod\limits_{i=1}^{m}(n_i)^2}, \qquad (3.30)$$

where α_2 is a constant.

It follows from (3.28) - (3.30) that

$$\sup_{\vec{T} \in G_0(\vec{n})} \left| q_{jk}(\vec{T}) - q_{jk}(\vec{n}^2) \right| \to 0, \quad \vec{n} \to \infty$$

with probability 1. Thus (3.27) implies the second part of the lemma.

Suppose that assumptions $1, 2, 3a), 4$ are true. Because of assumption 4 the matrix Q is positive definite. Lemma 15 implies that $Q(\vec{T}) \to Q$, $\prod\limits_{i=1}^{m} T_i \to \infty$

in probability. Then with probability, converging to 1 as $\prod_{i=1}^{m} T_i \to \infty$, the matrix $Q(\vec{T})$ is positive definite. Consequently, (3.24) implies that for every $\vec{T} \in \mathcal{R}^m$, $\vec{T} > 0$ and $\omega \in \Omega(\vec{T})$, $P\left(\Omega(\vec{T})\right) \to 1$, $\prod_{i=1}^{m} T_i \to \infty$, the function (3.23) has a single minimum point

$$\vec{\theta}\,(\vec{T}) = \vec{\theta}\,(\vec{T}, \omega) = \left(\vec{Q}\,(\vec{T})\right)^{-1} \frac{1}{\prod\limits_{i=1}^{m} T_i} \int\limits_{G(\vec{T})} x(\vec{t})\,\vec{y}\,(\vec{t})d\,\vec{t}\,. \qquad (3.31)$$

If also assumption 3b) holds then with probability 1, beginning from some numbers T_i, $i = \overline{1, m}$, depending on ω, the matrix $Q(\vec{T})$ is positive definite and there exists a single least squares estimate of the vector $\vec{\theta}$, defined by (3.31).

Theorem 26 *If assumptions* $1, 2, 3a), 3c), 4$ *hold then for any* $\varepsilon > 0$

$$P\left\{\omega \in \Omega(\vec{T}) : \|\vec{\theta}\,(\vec{T}, \omega) - \vec{\theta}\,\| > \varepsilon\right\} \to 0,$$

$$P\left\{\omega \in \Omega(\vec{T}) : |F_{\vec{T}}\left(\vec{\theta}\,(\vec{T}, \omega), \omega\right) - r(\vec{0})| > \varepsilon\right\} \to 0, \quad \prod_{i-1}^{m} T_i \to \infty.$$

If assumptions $1 - 4$ *hold then*

$$P\left\{\vec{\theta}\,(\vec{T}) \to \vec{\theta}, \quad F_{\vec{T}}\left(\vec{\theta}\,(\vec{T})\right) \to r(\vec{0}), \quad \vec{T} \to \infty\right\} = 1.$$

Proof. Let assumptions $1, 2, 3a), 3c), 4$ hold. For all $\vec{T} \in \mathcal{R}^m$, $\vec{T} > 0$ and $\omega \in \Omega(\vec{T})$

$$\vec{\theta}\,(\vec{T}) - \vec{\theta} = \left(Q(\vec{T})\right)^{-1} \frac{1}{\prod\limits_{i=1}^{m} T_i} \int\limits_{G(\vec{T})} \xi(\vec{t})\,\vec{y}\,(\vec{t})d\,\vec{t},$$

$$F_{\vec{T}}\left(\vec{\theta}\,(\vec{T})\right) =$$

$$= \frac{1}{\prod\limits_{i=1}^{m} T_i} \int\limits_{G(\vec{T})} \left(x(\vec{t}) - \left(\vec{y}(\vec{t}) \right)' \vec{\theta} + \left(\vec{y}(\vec{t}) \right)' \left(\vec{\theta} - \vec{\theta}(\vec{T}) \right) \right)^2 d\vec{t} =$$

$$= \frac{1}{\prod\limits_{i=1}^{m} T_i} \int\limits_{G(\vec{T})} \left(\xi(\vec{t}) \right)^2 d\vec{t} + \frac{2}{\prod\limits_{i=1}^{m} T_i} \int\limits_{G(\vec{T})} \xi(\vec{t}) \left(\vec{y}(\vec{t}) \right)' d\vec{t} \times$$

$$\times \left(\vec{\theta} - \vec{\theta}(\vec{T}) \right) + \left(\vec{\theta} - \vec{\theta}(\vec{T}) \right)' Q(\vec{T}) \left(\vec{\theta} - \vec{\theta}(\vec{T}) \right). \qquad (3.32)$$

In the same way as in Lemma 15 it can be shown that

$$\frac{1}{\prod\limits_{i=1}^{m} T_i} \int\limits_{G(\vec{T})} \xi(\vec{t}) \, \vec{y}(\vec{t}) d\vec{t} \to \vec{0}, \quad \frac{1}{\prod\limits_{i=1}^{m} T_i} \int\limits_{G(\vec{T})} \left(\xi(\vec{t}) \right)^2 d\vec{t} \to r(\vec{0}),$$

$$\prod\limits_{i=1}^{m} T_i \to \infty$$

in probability. Indeed by virtue of properties of Gaussian distributions [32], p.27, 28 for every $\vec{t}, \vec{s} \in \mathcal{R}^m$

$$E\left\{ \xi(\vec{t}) y_j(\vec{t}) \xi(\vec{s}) y_j(\vec{s}) \right\} =$$

$$= E\left\{ E\left(\xi(\vec{t}) y_j(\vec{t}) \xi(\vec{s}) y_j(\vec{s}) / \left(y_j(\vec{t}), y_j(\vec{s}) \right) \right) \right\} =$$

$$= r(\vec{t} - \vec{s}) \left(r_{jj}(\vec{t} - \vec{s}) + (a_j)^2 \right), \quad j = \overline{1, l};$$

$$E\left\{ \left(\left(\xi(\vec{t}) \right)^2 - r(\vec{0}) \right) \left(\left(\xi(\vec{s}) \right)^2 - r(\vec{0}) \right) \right\} =$$

$$= E\left\{ \left(\xi(\vec{t}) \right)^2 \left(\xi(\vec{s}) \right)^2 \right\} - \left(r(\vec{0}) \right)^2 = 2 \left(r(\vec{t} - \vec{s}) \right)^2.$$

That is why the statements similar to ones in Lemma 15 are applicable.

Lemma 15 implies that for every $\varepsilon > 0$

$$P\left\{\omega \in \Omega(\vec{T}) \; : \; \| \left(Q(\vec{T}, \omega)\right)^{-1} - Q^{-1} \| > \varepsilon \right\} \to 0, \quad \prod_{i=1}^{m} T_i \to \infty. \qquad (3.33)$$

From (3.32) and properties of probability we obtain the first part of the theorem.

The second part may be proved in the same way.

Let us note that

$$r(\vec{0}) = F(\vec{\theta}) = \min_{\vec{u} \in \mathcal{R}^l} F(\vec{u}),$$

where

$$F(\vec{u}) = E\left(x(\vec{0}) - \left(y(\vec{0})\right)' \vec{u}\right)^2, \quad \vec{u} \in \mathcal{R}^l.$$

Theorem 27 *If assumptions* $1, 2, 3a), 3c), 4$ *hold then for any* $\vec{v} \in \mathcal{R}^l$

$$\int_{G(\vec{T})} e^{i \vec{v}' \vec{\Delta}(\vec{T}, \omega)} \, dP \to e^{-\vec{v}' Q^{-1} H Q^{-1} \vec{v}/2}, \quad \vec{T} \to \infty,$$

where

$$\vec{\Delta}(\vec{T}) = \vec{\Delta}(\vec{T}, \omega) = \left(\prod_{i=1}^{m} T_i\right)^{1/2} \left(\vec{\theta}(\vec{T}, \omega) - \vec{\theta}\right),$$

$$H = \left(a_j \, a_k \int_{\mathcal{R}^m} r(\vec{t}) \, d\vec{t} + \int_{\mathcal{R}^m} r(\vec{t}) \, r_{jk}(\vec{t}) \, d\vec{t}\right)_{j,k=1}^{l}.$$

Proof. We have

$$\vec{\Delta}(\vec{T}) = \left(Q(\vec{T})\right)^{-1} \vec{\zeta}(\vec{T}), \quad \vec{T} > 0, \quad \omega \in \Omega(\vec{T}),$$

where

$$\vec{\zeta}(\vec{T}) = \frac{1}{\left(\prod\limits_{i=1}^{m} T_i\right)^{1/2}} \int\limits_{G(\vec{T})} \xi(\vec{t})\, \vec{y}(\vec{t}) d\,\vec{t}\,.$$

Then (3.33) and properties of probability imply that it is sufficient to prove that distribution of $\vec{\zeta}(\vec{T})$ converges weakly to $\mathcal{N}(\vec{0}, H)$ as $\vec{T} \to \infty$.

Fix $\vec{T} > 0$. Because of properties of Gaussian vectors and conditional expectations [32], $pp.27, 28$, [9], for any $n \in \mathcal{N}$ and $\vec{t}_i \in \mathcal{R}^m$, $i = \overline{1, n}$, the random vector $\left(\xi(\vec{t}_i)\right)_{i=1}^{n}$ has Gaussian conditional distribution with σ-algebra F. Consequently, conditional distribution $\vec{\zeta}(\vec{T})$ with F is Gaussian [32], $pp.27, 28$; [31], $p.49$; [100], with parameters [100]

$$E\left\{\vec{\zeta}(\vec{T}) / F\right\} = \vec{0},$$

$$E\left\{\vec{\zeta}(\vec{T})\left(\vec{\zeta}(\vec{T})\right)' / F\right\} =$$

$$= \frac{1}{\prod\limits_{i=1}^{m} T_i} \int\limits_{\left(G(\vec{T})\right)^2} E\left\{\xi(\vec{t})\xi(\vec{s}) / F\right\} \vec{y}(\vec{t})\left(\vec{y}(\vec{s})\right)' d\,\vec{t}\, d\,\vec{s} =$$

$$= \frac{1}{\prod\limits_{i=1}^{m} T_i} \int\limits_{\left(G(\vec{T})\right)^2} r(\vec{t} - \vec{s})\, \vec{y}(\vec{t})\left(\vec{y}(\vec{s})\right)' d\,\vec{t}\, d\,\vec{s} = H(\vec{T}) =$$

$$= \left(h_{jk}(\vec{T})\right)_{j,k=1}^{l},$$

where equalities are true with probability 1. Then we obtain that for every $\vec{v} \in \mathcal{R}^l$

$$E\, e^{i\,\vec{v}'\,\vec{\zeta}(\vec{T})} = E\left\{E\left(e^{i\,\vec{v}'\,\vec{\zeta}(\vec{T})} / F\right)\right\} = E\, e^{-\vec{v}'\, H(\vec{T})\, \vec{v}/2}. \tag{3.34}$$

Now we will show that

$$H(\vec{T}) \to H, \quad \vec{T} \to \infty \tag{3.35}$$

in probability.

Fix arbitrary $j, k \in \overline{1,l}$ and $\vec{T} > 0$. We have

$$h_{jk}(\vec{T}) = \frac{1}{\prod\limits_{i=1}^{m} T_i \left(G(\vec{T})\right)^2} \int r(\vec{t} - \vec{s}) \, y_j(\vec{t}) y_k(\vec{s}) \, d\vec{t} \, d\vec{s} =$$

$$= D_{jk}^1(\vec{T}) + D_{jk}^2(\vec{T}) + D_{jk}^3(\vec{T}) + D_{jk}^4(\vec{T}), \tag{3.36}$$

where

$$D_{jk}^1(\vec{T}) = \frac{1}{\prod\limits_{i=1}^{m} T_i \left(G(\vec{T})\right)^2} \int r(\vec{t} - \vec{s}) \, \vec{y}_j^{\,0}(\vec{t}) \, \vec{y}_k^{\,0}(\vec{s}) \, d\vec{t} \, d\vec{s},$$

$$D_{jk}^2(\vec{T}) = \frac{a_k}{\prod\limits_{i=1}^{m} T_i \left(G(\vec{T})\right)^2} \int r(\vec{t} - \vec{s}) \, \vec{y}_j^{\,0}(\vec{t}) \, d\vec{t} \, d\vec{s},$$

$$D_{jk}^3(\vec{T}) = \frac{a_j}{\prod\limits_{i=1}^{m} T_i \left(G(\vec{T})\right)^2} \int r(\vec{t} - \vec{s}) \, \vec{y}_k^{\,0}(\vec{s}) \, d\vec{t} \, d\vec{s},$$

$$D_{jk}^4(\vec{T}) = \frac{a_j \, a_k}{\prod\limits_{i=1}^{m} T_i \left(G(\vec{T})\right)^2} \int r(\vec{t} - \vec{s}) \, d\vec{t} \, d\vec{s}.$$

For all $n \in \mathcal{N}$ denote

$$\chi_A(\vec{d}) = \begin{cases} 1, & \vec{d} \in A \\ \\ 0, & \vec{d} \notin A \end{cases}, \quad A \subset \mathcal{R}^n, \ \vec{d} \in \mathcal{R}^n.$$

After changing variables and using Fubini theorem we obtain

$$
D^4_{jk}(\vec{T}) = \frac{a_j\, a_k}{\prod\limits_{i=1}^{m} T_i} \int\limits_{G(\vec{T})} \left(\int\limits_{\Pi[-\vec{s},\,\vec{T}-\vec{s}\,]} r(\vec{t})\, d\,\vec{t} \right) d\,\vec{s} =
$$

$$
= \frac{a_j\, a_k}{\prod\limits_{i=1}^{m} T_i} \int\limits_{\Pi[-\vec{T},\,\vec{T}\,]} \left(r(\vec{t}) \int\limits_{G^1(\vec{t},\vec{T})} d\,\vec{s} \right) d\,\vec{t},
$$

where $G^1(\vec{t},\vec{T}) = \left\{ \vec{s} = (\vec{s}_i)_{i=1}^{m} \in \mathcal{R}^m \;:\; -t_i\, \chi_{\mathcal{R}_-}(t_i) \le s_i \le T_i - t_i\, \chi_{\mathcal{R}_+}(t_i), \right.$ $\left. i = \overline{1,m} \right\}$, $\vec{t} = (t_i)_{i=1}^{m} \in \Pi[-\vec{T},\vec{T}\,]$; $\mathcal{R}_- =]-\infty, 0]$.

Then

$$
D^4_{jk}(\vec{T}) = \frac{a_j\, a_k}{\prod\limits_{i=1}^{m} T_i} \int\limits_{\Pi[-\vec{T},\,\vec{T}\,]} r(\vec{t}) \prod_{i=1}^{m} \left(T_i - |t_i| \right) d\,\vec{t} =
$$

$$
= a_j\, a_k \int\limits_{\mathcal{R}^m} \chi_{\Pi[-\vec{T},\,\vec{T}\,]}(t)\, r(\vec{t}) \prod_{i=1}^{m} \left(1 - |t_i|/T_i \right) d\,\vec{t} \; .
$$

By virtue of Lebesgue theorem of limit transition we obtain

$$
\lim_{\vec{T}\to\infty} D^4_{jk}(\vec{T}) = a_j\, a_k \int\limits_{\mathcal{R}^m} r(\vec{t})\, d\,\vec{t} \; . \tag{3.37}
$$

The following relationship is proved in the same way:

$$
\frac{1}{\prod\limits_{i=1}^{m} T_i \left(G(\vec{T}) \right)^2} \int r(\vec{t} - \vec{s})\, r_{jk}(\vec{t} - \vec{s})\, d\,\vec{t}\, d\,\vec{s} \longrightarrow \int\limits_{\mathcal{R}^m} r(\vec{t})\, r_{jk}(\vec{t})\, d\,\vec{t},
$$

$$
\vec{T} \to \infty.
$$

For every $\vec{T} > 0$

$$E\left(D_{jk}^1(\vec{T}) - \frac{1}{\prod\limits_{i=1}^{m} T_i} \int\limits_{\left(G(\vec{T})\right)^2} r(\vec{t}-\vec{s})\, r_{jk}(\vec{t}-\vec{s})\, d\vec{t}\, d\vec{s}\right)^2 =$$

$$= \frac{1}{\prod\limits_{i=1}^{m}(T_i)^2} \int\limits_{\left(G(\vec{T})\right)^4} r(\vec{t}-\vec{s})\, r(\vec{t}^1-\vec{s}^1)\, E\left\{\left(y_j^0(\vec{t})\, y_k^0(\vec{s})-\right.\right.$$

$$\left.-r_{jk}(\vec{t}-\vec{s})\right)\left(y_j^0(\vec{t}^1)\, y_k^0(\vec{s}^1)-r_{jk}(\vec{t}^1-\vec{s}^1)\right)\bigg\}\, d\vec{t}\, d\vec{s}\, d\vec{t}^1\, d\vec{s}^1 =$$

$$= \frac{1}{\prod\limits_{i=1}^{m}(T_i)^2} \int\limits_{\left(G(\vec{T})\right)^4} r(\vec{t}-\vec{s})\, r(\vec{t}^1-\vec{s}^1)\left(r_{jj}(\vec{t}-\vec{t}^1)\, r_{kk}(\vec{s}-\vec{s}^1)+\right.$$

$$\left.+r_{jk}(\vec{t}-\vec{s}^1)\, r_{kj}(\vec{s}-\vec{t}^1)\right)d\vec{t}\, d\vec{s}\, d\vec{t}^1\, d\vec{s}^1 \le \frac{r(\vec{0})}{\prod\limits_{i=1}^{m}(T_i)^2} \times$$

$$\times \int\limits_{\left(G(\vec{T})\right)^2} |r(\vec{t}^1-\vec{s}^1)|\left(\int\limits_{\Pi[-\vec{t}^1,\vec{T}-\vec{t}^1]\times\Pi[-\vec{s}^1,\vec{T}-\vec{s}^1]} |r_{jj}(\vec{t})\, r_{kk}(\vec{s})|\, d\vec{t}\, d\vec{s}\, +\right.$$

$$+ \int\limits_{\Pi[-\vec{s}^1,\vec{T}-\vec{s}^1]\times\Pi[-\vec{t}^1,\vec{T}-\vec{t}^1]} |r_{jk}(\vec{t})\, r_{kj}(\vec{s})|\, d\vec{t}\, d\vec{s}\Bigg)\, d\vec{t}^1\, d\vec{s}^1 \le$$

$$\le \frac{r(\vec{0})\,(c_{jj}c_{kk}+(c_{jk})^2)}{\prod\limits_{i=1}^{m}(T_i)^2} \int\limits_{G(\vec{T})}\left(\int\limits_{\Pi[-\vec{s}^1,\vec{T}-\vec{s}^1]} |r(\vec{t})|\, d\vec{t}\right)d\vec{s}^1 \le$$

$$\leq \frac{r(\vec{0})\,(c_{jj}c_{kk} + (c_{jk})^2\,c)}{\prod\limits_{i=1}^{m} T_i}.$$

Consequently,

$$D^1_{jk}(\vec{T}) \xrightarrow{P} \int\limits_{\mathcal{R}^m} r(\vec{t})\,r_{jk}(\vec{t})\,d\,\vec{t}, \quad \vec{T} \to \infty. \tag{3.38}$$

Let us show that

$$D^2_{jk}(\vec{T}) \xrightarrow{P} 0, \quad \vec{T} \to \infty. \tag{3.39}$$

For any $\vec{T} > 0$

$$E\left(D^2_{jk}(\vec{T})\right)^2 = \frac{(a_k)^2}{\prod\limits_{i=1}^{m}(T_i)^2 \left(G(\vec{T})\right)^4} \int r(\vec{t} - \vec{s})\, r(\vec{t}^{\,1} - \vec{s}^{\,1}) \times$$

$$\times\; r_{jj}(\vec{t} - \vec{t}^{\,1})\,d\,\vec{t}\;d\,\vec{s}\;d\,\vec{t}^{\,1}\,d\,\vec{s}^{\,1} = \frac{(a_k)^2}{\prod\limits_{i=1}^{m}(T_i)^2 \left(G(\vec{T})\right)^2} \int \left(r_{jj}(\vec{t} - \vec{t}^{\,1}) \times \right.$$

$$\times\; \left(\int\limits_{\Pi[-\vec{t}^{\,1},\vec{T}-\vec{t}^{\,1}]\times\Pi[-\vec{s}^{\,1},\vec{T}-\vec{s}^{\,1}]} r(\vec{s})\,r(\vec{s}^{\,1})\,d\,\vec{s}\;d\,\vec{s}^{\,1} \right) d\,\vec{t}\;d\,\vec{t}^{\,1} \leq$$

$$\leq \frac{c^2\,(a_k)^2}{\prod\limits_{i=1}^{m}(T_i)^2}\int\limits_{G(\vec{T})} \left(\int\limits_{\Pi[-\vec{t}^{\,1},\vec{T}-\vec{t}^{\,1}]} |r_{jj}(\vec{t})|\,d\,\vec{t} \right) d\,\vec{t}^{\,1} \leq \frac{c^2\,c_{jj}\,(a_k)^2}{\prod\limits_{i=1}^{m} T_i},$$

that implies (3.39).

Then for all $\vec{T}$

$$D^3_{jk}(\vec{T}) = D^2_{kj}(\vec{T}) \xrightarrow{P} 0, \quad \vec{T} \to \infty. \tag{3.40}$$

Now (3.35) follows from (3.36) - (3.40).

Because correlation functions are non-negative definite, then for all $\vec{T} > 0$, $\vec{v} \in \mathcal{R}^l$ we have $\vec{v}' \, H(\vec{T}) \, \vec{v} \geq 0$ with probability 1 . That is why, passing to limit under the expectation sign by virtue of Lebesgue theorem, we obtain from (3.34), (3.35) that for any $\vec{v} \in \mathcal{R}^l$

$$E \, e^{i \, \vec{v}' \, \vec{\zeta} \, (\vec{T})} = E \, e^{- \, \vec{v}' \, H(\vec{T}) \, \vec{v}/2} \to E \, e^{- \, \vec{v}' \, H \vec{v}/2}, \quad \vec{T} \to \infty.$$

This implies the proof of the theorem.

Consider the partial case of our problem. Let assumptions $1, 2$ be fulfilled. We have observations

$$\left\{ \left(x(\vec{t}), \vec{y} \, (\vec{t}) \right), \; \vec{t} \in [0, T]^m \right\},$$

where $T > 0$ and the vector $\vec{\theta}$ is to be estimated.

Denote

$$Q(T) = \frac{1}{T^m} \int\limits_{[0,T]^m} \vec{y} \, (\vec{t}) \left(\vec{y} \, (\vec{t}) \right)' d \, \vec{t}, \quad T > 0.$$

Lemma 16 *If assumptions $2, 3a$) hold then*

$$P \left\{ Q(T) \to Q, \quad T \to \infty \right\} = 1,$$

where matrix Q is from Lemma 15.

Proof. Fix $j, k \in \{\overline{1, l}\}$ and denote $q_{jk}(T)$ and q_{jk} elements of matrices $Q(T)$ and Q. As it was shown in Lemma 15 for any $T > 0$

$$E \left(q_{jk}(T) - q_{jk} \right)^2 \leq \frac{\alpha_{jk}}{T^m},$$

where α_{jk} is the constant. That is why using the relation $m \geq 2$ and Borel-Cantelli lemma, for $n \in \mathcal{N}$

$$P \left\{ q_{jk}(n) \to q_{jk}, \quad n \to \infty \right\} = 1.$$

Then,

$$\sup_{T \in \mathcal{R}: T > T_0} \left| q_{jk}(T) - q_{jk} \right| \leq \sup_{n \in \mathcal{N}: n > T_0 - 1} \; \sup_{T \in [n, n+1[} \left| q_{jk}(T) - q_{jk}(n) \right| +$$

$$+ \sup_{n \in \mathcal{N}: n > T_0 - 1} \left| q_{jk}(n) - q_{jk} \right|, \quad T_0 \geq 1 \quad (3.41)$$

with probability 1.

For any $n \in \mathcal{N}$ and $T \in [n, n+1[$

$$\left| q_{jk}(T) - q_{jk}(n) \right| \leq \frac{1}{n^m} \left| \int\limits_{[0,T]^m \setminus [0,n]^m} y_j(\vec{t}) y_k(\vec{t}) d\vec{t} \right| +$$

$$+ \quad (1 - 1/(1+1/n)^m) \left| q_{jk}(n) \right| \qquad (3.42)$$

with probability 1. After applying Cauchy-Buniakowski inequality, we have

$$E \left(\sup_{T \in [n,n+1[} \frac{1}{n^m} \left| \int\limits_{[0,T]^m \setminus [0,n]^m} y_j(\vec{t}) y_k(\vec{t}) d\vec{t} \right| \right)^2 \leq$$

$$\leq \frac{1}{n^{2m}} E \sup_{T \in [n,n+1[} \left((T^m - n^m) \int\limits_{[0,T]^m \setminus [0,n]^m} \left(y_j(\vec{t}) y_k(\vec{t}) \right)^2 d\vec{t} \right) =$$

$$= \frac{\left((n+1)^m - n^m \right)^2 \left(\alpha_{jk}(\vec{0}) + (q_{jk})^2 \right)}{n^{2m}} \leq \frac{\alpha}{n^2}, \quad n \in \mathcal{N},$$

where the function $\alpha_{jk}(\vec{t})$ is from (3.25), α is the constant.

Consequently, (3.41) implies that

$$P \left\{ \sup_{T \in [n,n+1[} \left| q_{jk}(T) - q_{jk}(n) \right| \to 0, \quad n \to \infty \right\} = 1.$$

Now the proof of the lemma follows from (3.43).

Because of Lemma 16 under assumptions $2, 3a), 4$ beginning from some T which depends on ω, there exists with probability 1 a single minimum point

$$\vec{\theta}(T) = \left(Q(T) \right)^{-1} \frac{1}{T^m} \int\limits_{[0,T]^m} x(\vec{t}) \, \vec{y}(\vec{t}) d\vec{t}$$

of the function

$$F_T(\vec{u}) = \frac{1}{T^m} \int\limits_{[0,T]^m} \left(x(\vec{t}) - \left(y(\vec{t}) \right)' \vec{u} \right)^2 d\vec{t}, \quad \vec{u} \in \mathcal{R}^l.$$

Theorem 28 *Let assumptions* $1, 2, 3a), 3c), 4$ *be fulfilled. Then*

$$P\left\{ \vec{\theta}\,(T) \to \vec{\theta}, \quad F_T\left(\vec{\theta}\,(T) \right) \to r(\vec{0}); \quad T \to \infty \right\} = 1.$$

The proof of the theorem is similar to that one for Theorem 26, and Lemma 16 was applied.

Now we will investigate our regression model for the case when the field $\left(x(\vec{t}), \vec{y}\,(\vec{t}) \right)$ is observed on a ball in $\mathcal{R}^m$.

Denote

$$S_R^0(\vec{b}) = \left\{ \vec{t} \in \mathcal{R}^m \; : \; \| \vec{t} - \vec{b} \| \le R \right\}, \quad \vec{b} \in \mathcal{R}^m;$$

$$S_R = S_R^0(\vec{0}); \quad R \in \mathcal{R}_+.$$

Suppose that assumptions $1, 2$ are hold. We need to estimate the vector $\vec{\theta}$ by observations

$$\left\{ \left(x(\vec{t}), \vec{y}\,(\vec{t}) \right), \quad \vec{t} \in S_R \right\}, \quad R > 0.$$

Introduce for all $R > 0$

$$F_R(\vec{u}) = \frac{1}{V_R} \int\limits_{S_R} \left(x(\vec{t}) - \left(\vec{y}\,(\vec{t}) \right)' \vec{u} \right)^2 d\vec{t}, \quad \vec{u} \in \mathcal{R}^l, \qquad (3.43)$$

where

$$V_R = \int\limits_{S_R} d\vec{t} = \frac{2\pi^{m/2} R^m}{m\Gamma(m/2)}; \quad \Gamma(p) = \int\limits_{0}^{+\infty} z^{p-1} e^{-z}\, dz, \quad p \in\,]0, +\infty[$$

is the special function [102] (*pp*.111, 112, 193). The function (3.43) is twice differentiable on $\mathcal{R}^l$ as well as function (3.23), and their partial derivatives differ only in the coefficient before the integral and the set of integration. That is why the necessary and sufficient conditions of minimum for functions (3.23) and (3.43) are similar.

Denote

$$Q(R) = Q(R,\omega) = \frac{1}{V_R} \int_{S_R} \vec{y}\,(\vec{t})\left(\vec{y}\,(\vec{t})\right)' d\,\vec{t}, \quad R > 0.$$

Lemma 17 *Suppose that assumptions 2, 3a) are satisfied. Then*

$$P\Big\{Q(R) \to Q, \quad R \to \infty\Big\} = 1,$$

where the matrix Q is from Lemma 15.

Proof. Fix $j, k \in \{\overline{1,l}\}$. Denote by $q_{jk}(R)$ and q_{jk} the elements of matrices $Q(R)$ and Q. As in Lemma 15 for every $R > 0$ we have

$$E\left(q_{jk}(R) - q_{jk}\right)^2 = \frac{1}{(V_R)^2} \int_{(S_R)^2} \alpha_{jk}(\vec{t} - \vec{s})\, d\,\vec{t}\, d\,\vec{s} =$$

$$= \frac{1}{(V_R)^2} \int_{S_R} \left(\int_{S_R^0(-\vec{s})} \alpha_{jk}(\vec{t})\, d\,\vec{t} \right) d\,\vec{s} \le \frac{\alpha_{jk}}{V_R} = \frac{\alpha_{jk}^1}{R^m},$$

where the function $\alpha_{jk}(\vec{t})$ and constant α_{jk} are defined in Lemma 15, and

$$\alpha_{jk}^1 = \frac{\alpha_{jk}\, m\, \Gamma(m/2)}{2\, \pi^{m/2}}.$$

Consequently, from the relation $m \ge 2$ we obtain for $n \in \mathcal{N}$

$$P\Big\{q_{jk}(n) \to q_{jk}, \quad n \to \infty\Big\} = 1. \qquad (3.44)$$

For all $n \in \mathcal{N}$ and $R \in [n, n+1[$

$$\left| q_{jk}(R) - q_{jk}(n) \right| \le$$

$$\le \frac{1}{V_R} \left| \int\limits_{S_R \setminus S_n} y_j(\vec{t}) \, y_k(\vec{t}) \, d\vec{t} \right| + \left| \left(1/(V_R/V_n) - 1 \right) q_{jk}(n) \right| \le$$

$$\le \frac{1}{V_n} \left| \int\limits_{S_R \setminus S_n} y_j(\vec{t}) \, y_k(\vec{t}) \, d\vec{t} \right| + \left| \left(1 - 1/(1 + 1/n)^m \right) q_{jk}(n) \right|$$

with probability 1. Then

$$E \left(\sup_{R \in [n, n+1[} \frac{1}{V_n} \left| \int\limits_{S_R \setminus S_n} y_j(\vec{t}) \, y_k(\vec{t}) \, d\vec{t} \right| \right)^2 \le$$

$$\le \frac{1}{(V_n)^2} E \sup_{R \in [n, n+1[} \left((V_R - V_n) \int\limits_{S_R \setminus S_n} \left(y_j(\vec{t}) \, y_k(\vec{t}) \right)^2 d\vec{t} \right) =$$

$$= \frac{(V_{n+1} - V_n)^2 \left(\alpha_{jk}(\vec{0}) + (q_{jk})^2 \right)}{(V_n)^2} \le \frac{\alpha}{n^2}, \quad n \in \mathcal{N},$$

where α is the constant. Then,

$$P \left\{ \sup_{R \in [n, n+1[} |q_{jk}(R) - q_{jk}(n)| \to 0, \quad n \to \infty \right\} = 1. \qquad (3.45)$$

As in Lemma 16, the lemma follows from (3.41), (3.45).

If assumptions $2, 3a), 4$ hold then because of Lemma 17, beginning from some R, which depends on ω, there exists with probability 1 a single minimum point for function (3.43):

$$\vec{\theta}(R) = \vec{\theta}(R, \omega) = \left(\vec{Q}(R) \right)^{-1} \frac{1}{V_R} \int\limits_{S_R} x(\vec{t}) \, \vec{y}(\vec{t}) \, d\vec{t} .$$

Theorem 29 *If assumptions* 2, 3a), 3c), 4 *hold then with probability 1*

$$\vec{\theta}\,(R) \to \vec{\theta}, \quad F_R\left(\vec{\theta}\,(R)\right) \to r(\vec{0}); \quad R \to \infty.$$

The proof of the theorem is similar to that one in Theorem 26.

Denote

$$\Omega(R) = \left\{\omega \in \Omega : \text{ the function } \vec{\theta}\,(R,\omega) \text{ is defined}\right\};$$

$$\vec{\Delta}\,(R,\omega) = \sqrt{V_R}\left(\vec{\theta}\,(R,\omega) - \vec{\theta}\right), \quad \omega \in \Omega(R); \quad R > 0.$$

It is evident that $P\left(\Omega(R)\right) \to 1$, as $R \to \infty$.

Theorem 30 *If assumptions* 1, 2, 3a), 3c), 4 *hold then for any* $\vec{v} \in \mathcal{R}^l$

$$\int\limits_{\Omega(R)} e^{i\,\vec{v}'\,\vec{\Delta}\,(R,\omega)}\,dP \to e^{-\,\vec{v}'\,Q^{-1}\,H\,Q^{-1}\,\vec{v}/2}, \quad R \to \infty,$$

where the matrix H is defined in Theorem 27.

Proof. As in Theorem 27 it is sufficient to show that the matrix

$$H(R) = (h_{jk}(R))_{j,k=1}^l = \frac{1}{V_R} \int\limits_{(S_R)^2} r(\vec{t} - \vec{s})\,\vec{y}\,(\vec{t})\left(\vec{y}\,(\vec{s})\right)'\,d\vec{t}\,d\vec{s}$$

converges in probability to H as $R \to \infty$.

Fix $j, k \in \{\overline{1, l}\}$. For all $R > 0$

$$D_{jk}(R) \quad = \quad \frac{a_j\,a_k}{V_R} \int\limits_{(S_R)^2} r(\vec{t} - \vec{s})\,d\vec{t}\,d\vec{s} =$$

$$= \quad \frac{a_j\,a_k}{V_R} \int\limits_{S_R} \left(\int\limits_{S_R^0(-\vec{s})} r(\vec{t})\,d\vec{t}\right)\,d\vec{s} =$$

$$= \frac{a_j a_k}{V_R} \int\limits_{S_{2R}} \left(r(\vec{t}) \int\limits_{S_R \cap S_R^0(-\vec{t})} d\vec{s} \right) d\vec{t} =$$

$$= a_j a_k \int\limits_{\mathcal{R}^m} \chi_{S_{2R}}(\vec{t}) r(\vec{t}) U_R(\vec{t}) d\vec{t},$$

where

$$U_R(\vec{t}) = \frac{1}{V_R} \int\limits_{S_R \cap S_R^0(-\vec{t})} d\vec{s}, \quad \vec{t} \in \mathcal{R}^m,$$

$$\chi_{S_{2R}}(\vec{t}) = \begin{cases} 1, & \vec{t} \in S_{2R} \\ 0, & \vec{t} \in\!\!\!\!\!/\, S_{2R} \end{cases}.$$

Fix arbitrary $\vec{t} \in \mathcal{R}^m$. We will show that

$$U_R(\vec{t}) \to 1, \quad R \to \infty.$$

Fix $R > \| \vec{t} \|$. The inequality $\| \vec{s} \| \le R - \| \vec{t} \|$ implies that $\| \vec{s} + \vec{t} \| \le R$. Then we have

$$S_{R-\| \vec{t} \|} \subset S_R \cap S_R^0(-\vec{t}).$$

Hence

$$1 \ge U_R(\vec{t}) \ge \frac{1}{V_R} \int\limits_{S_{R-\| \vec{t} \|}} d\vec{s} = \frac{(R - \| \vec{t} \|)^m}{R^m} = \left(1 - \frac{\| \vec{t} \|}{R} \right)^m \to 1,$$

$$R \to \infty.$$

We obtain that for every $\vec{t} \in \mathcal{R}^m$

$$\chi_{S_{2R}}(\vec{t}) r(\vec{t}) U_R(\vec{t}) \to r(\vec{t}), \quad R \to \infty.$$

After applying Lebesgue theorem of limit transition we have

$$D_{jk}(R) \to a_j\, a_k \int_{\mathcal{R}^m} r(\vec{t})\, d\,\vec{t}, \quad R \to \infty.$$

Then the proof is similar to that one in Theorem 27. The proof is complete.

3.3 NONSTATIONARY REGRESSION MODEL FOR GAUSSIAN FIELD

Now we shall investigate some properties of the least squares estimate for non-homogeneous Gaussian random field. Let $\{a_1, \ldots, a_m\}$ be the fixed set of real continuous functions on $[0, \infty) \times [0, \infty)$, satisfying the conditions presented below. To formulate the conditions we put

$$\vec{a}' = (a_1, a_2, \ldots, a_m), \quad q_k^2(u, v) = \iint_{G(u,v)} a_k^2(s, t)\, ds\, dt, \quad u > 0, \quad v > 0,$$

where

$$G(u, v) = G(0, u; \; 0, v),$$

$$G(a, b; \; c, d) = \Big\{ (s, t) \; : \; a \le s \le b, \quad c \le t \le d \Big\};$$

$$Q(u, v) = \Big(q_k^2(u, v)\, \delta_{jk} \Big)_{j,k=1}^m,$$

$$\vec{a}\,(u, v) = Q^{-\frac{1}{2}}(u, v)\, \vec{a} = \Big(a_1^{(u,v)}, a_2^{(u,v)}, \ldots, a_m^{(u,v)} \Big).$$

1a. For some number L_1, all $1 \le k \le m$, and $u \ge S_0 > 0$, $v \ge T_0 > 0$, we have

$$q_k^{-1}(u, v) \max_{\big((s,t) \in G(u,v)\big)} \Big| a_k(s, t) \Big| \le L_1 (uv)^{-\frac{1}{2}}.$$

1b. For some number L_2, for any real s, t and any $u \geq S_0 > 0$, $v \geq T_0 > 0$, the inequalities

$$\left\| \iint\limits_{G(u,v)} \vec{a}^{(u,v)} \left(\tilde{s} + |s|, \tilde{t} + |t| \right) \vec{a}^{(u,v)} (\tilde{s}, \tilde{t})' \, d\tilde{s} \, d\tilde{t} - R(s,t) \right\| \leq \frac{L_2}{\sqrt{uv}},$$

$$\left\| \iint\limits_{G(u,v)} \vec{a}^{(u,v)} \left(\tilde{s} + |s|, \tilde{t} \right) \vec{a}^{(u,v)} \left(\tilde{s}, \tilde{t} + |t| \right)' \, d\tilde{s} \, d\tilde{t} - R(s,t) \right\| \leq \frac{L_2}{\sqrt{uv}},$$

hold with some symmetric matrix

$$R(s,t) = (R_{jk}(s,t))_{jk=1}^{m} = (R_{kj}(s,t))_{jk=1}^{m}$$

having elements continuous in $\mathcal{R}^2$ and for which the matrix $R(0,0)$ is positive.

1c. For some number $L_3 > 0$ and all $1 \leq k \leq m$, $u > S_0$, $v > T_0$ the following inequalities are valid:

$$\frac{q_k^2(2u,v)}{q_k^2(u,v)} \leq L_3 u, \quad \frac{q_k^2(u,2v)}{q_k^2(u,v)} \leq L_3 u, \quad \frac{q_k^2(2u,2v)}{q_k^2(u,v)} \leq L_3 u v.$$

1d. For any vector $\vec{a} \in \mathcal{R}^m$ with $\| \vec{a} \| \neq 0$ the function $\vec{a}' \vec{a} \, (s,t)$, $(s,t) \in \mathcal{R}^2$, cannot be equal to 0 on a set of positive Lebesgue measure in $\mathcal{R}^2$.

Let $\left\{ x(s,t), \, y(s,t); \, (s,t) \in \mathcal{R}^2 \right\}$ be two-dimensional Gaussian field with real components and continuous with probability 1 sample functions. Denote by F the σ-algebra generated by the random variables $y(s,t); \, (s,t) \in \mathcal{R}^2$.

2a. The random field $\{y(s,t); \, (s,t) \in \mathcal{R}^2\}$ is homogeneous with the mean $E \, y(s,t) = a$ and the correlation function

$$r_1(s,t) = E \left\{ y_0(s,t) \, y_0(0,0) \right\}, \quad y_0(s,t) = y(s,t) - a,$$

such that following integrals are finite:

$$c_{10} = \int\limits_{-\infty}^{\infty} \left| r_1(s,0) \right| ds, \quad c_{01} = \int\limits_{-\infty}^{\infty} \left| r_1(0,t) \right| dt,$$

$$c_{11} = \int\limits_{-\infty}^{\infty} \int\limits_{-\infty}^{\infty} \left| r_1(s,t) \right| ds\, dt.$$

2b. The function r_1 satisfies the following condition of strictly positive definiteness: for any $u, v > 0$ a function $g \in L_2\left(G(u,v)\right)$ for which

$$\iint\limits_{G(u,v)} \iint\limits_{G(u,v)} r_1\left(s - \tilde{s},\ t - \tilde{t}\right) g(s,t)\, g(\tilde{s},\tilde{t})\, ds\, dt\, d\tilde{s}\, d\tilde{t} = 0$$

equals to 0 almost everywhere on $G(u,v)$ in Lebesgue measure on $\mathcal{R}^2$.

3. For some fixed vector $\vec{\theta} \in \mathcal{R}^m$ and any $(s,t) \in \mathcal{R}^2$ we have with probability 1

$$E\left\{x(s,t)/F\right\} = \vec{a}\,(s,t)'\,\vec{\theta}\,y(s,t),$$

$$E\left\{z\left(s + \tilde{s},\ t + \tilde{t}\right) z(\tilde{s},\tilde{t})/F\right\} = r(s,t),$$

where

$$z(s,t) = x(s,t) - E\left\{x(s,t)/F\right\}.$$

The function r is assumed to be nonrandom and such that

$$c_0 = \int\limits_{-\infty}^{\infty} \int\limits_{-\infty}^{\infty} \left| r(u,v) \right| du\, dv < \infty.$$

Let us consider as an estimator of the vector $\vec{\theta}$ from the observation $\left\{x(u,v),\ y(u,v); \in G(s,t)\right\}$ the simplest estimator $\vec{\theta}\,(S,T)$ obtained by the method of least squares. This estimator has the form

$$\vec{\theta}\,(S,T) = \Gamma^{-1}(S,T) \iint\limits_{G(u,v)} x(u,v)\,y(u,v)\,\vec{a}\,(u,v)\,du\,dv$$

with the matrix $\Gamma(S,T)$ defined by

$$\Gamma(S,T) = \iint\limits_{G(u,v)} \vec{a}\,(u,v)\,\vec{a}\,(u,v)'\,y^2(u,v)\,du\,dv.$$

Under conditions 1d and 2a the matrix $\Gamma(S,T)$ is, for any $T > 0$, positive definite with probability 1. Thus, $\Gamma^{-1}(S,T)$ is defined with probability 1, and consequently so is $\vec{\theta}\,(S,T)$.

First we prove the lemma which describes the limit behavior of the elements of $\Gamma(S,T)$.

Lemma 18 *Assume that conditions 1a, 1b and 2a are satisfied. Then the relation*

$$P\left\{\lim_{S\to\infty,\,T\to\infty} Q^{-\frac{1}{2}}(S,T)\,\Gamma(S,T)\,Q^{-\frac{1}{2}}(S,T) = \Gamma\right\} = 1$$

holds with positive definite matrix

$$\Gamma = (\gamma_{jk})_{jk=1}^{m} = \left[a^2 + r_1(0,0)\right]\,R(0,0).$$

Proof. For $u,v > 0$ let

$$\gamma_{jk}(u,v) = \iint\limits_{G(u,v)} a_j^{(u,v)}(s,t)\,a_k^{(u,v)}(s,t)\,y^2(s,t)\,ds\,dt$$

be the element with indices j,k of the matrix

$$Q^{-\frac{1}{2}}(u,v)\,\Gamma(u,v)\,Q^{-\frac{1}{2}}(u,v).$$

On the basis of conditions 1*a*, 1*b* and 2*a* it is easy to obtain

$$E\left[\gamma_{jk}(u,v) - \gamma_{jk}\right]^2 \le 2\left[E\,\gamma_{jk}(u,v) - \gamma_{jk}\right]^2 + 2E\left[\gamma_{jk}(u,v) - \right.$$

$$- \left. E\,\gamma_{jk}(u,v)\right]^2 = 2\left(E\left\{y^2(0,0)\right\}\right)^2\left[\iint\limits_{G(u,v)} a_j^{(u,v)}(s,t)\,a_k^{(u,v)}(s,t)\,ds\,dt - \right.$$

$$- \left. R_{jk}\right]^2 + 4\iint\limits_{G(u,v)}\iint\limits_{G(u,v)} a_j^{(u,v)}(s,t)\,a_k^{(u,v)}(s,t)\,a_j^{(u,v)}(\tilde{s},\tilde{t})\,a_k^{(u,v)}(\tilde{s},\tilde{t}) \times$$

$$\times \left[r_1^2(s-\tilde{s},\ t-\tilde{t}) + 2a^2 r_1(s-\tilde{s},\ t-\tilde{t})\right]\,ds\,dt\,d\tilde{s}\,d\tilde{t} \le$$

$$\le L'\cdot(u,v)^{-1} + L''\cdot(u,v)^{-2}\iint\limits_{G(u,v)}\iint\limits_{G(u,v)}\left[r_1^2(s-\tilde{s},\ t-\tilde{t}) + \right.$$

$$+ \left. 2a^2 r_1(s-\tilde{s},\ t-\tilde{t})\right]\,ds\,dt\,d\tilde{s}\,d\tilde{t} \le L(u,v)^{-1}, \tag{3.46}$$

where

$$L' = 2L_2^2\left[E\left\{y^2(0,0)\right\}\right]^2, \quad L'' = 4L_1^4, \quad L = L' + c_{11}L''\left[r_1(0,0) + 2a^2\right].$$

Using the same conditions, we come analogously to

$$E\left\{\max_{((u,v)\in G(u_1,u_2;\ v_1,v_2))} \times \right.$$

$$\times \left[\iint\limits_{G(0,u_1;\ v_1,v_2)} a_j^{(u,v)}(s,t)\,a_k^{(u,v)}(s,t)\left(y^2(s,t) - E\,y^2(0,0)\right)\,ds\,dt\right]^2\left.\right\} \le$$

$$\le E\left\{\left[\int\limits_{v_1}^{v_2}\left|\int\limits_0^{u_2} a_j^{(u_1,v_1)}(s,t)\,a_k^{(u_1,v_1)}(s,t)\left(y^2(s,t) - E\,y^2(0,0)\right)\,ds\right|\,dt\right]^2\right\} \le$$

$$\leq \ (v_2 - v_1) \int\limits_{v_1}^{v_2} E \left\{ \int\limits_{0}^{u_1} a_j^{(u_1,v_1)}(s,t)\, a_k^{(u_1,v_1)}(s,t) \left(y^2(s,t) - E\, y^2(0,0) \right) ds \right\}^2 dt \leq$$

$$\leq \ \frac{2L_1^4(v_2 - v_1)^2}{u_1^2 v_1^2} \iint\limits_{G(u_1,u_2)} \left[r_1^2\,(s - t,\, 0) + 2a^2 r_1\,(s - t,\, 0) \right] ds\, dt \leq$$

$$\leq \ 2\, L_1^4 \left[r_1(0,0) + 2a^2 \right] c_{10}\, \frac{(v_2 - v_1)^2}{u_1\, v_1^2}, \tag{3.47}$$

$$E \left\{ \max_{\left((u,v) \in G(u_1,u_2;\ v_1,v_2) \right)} \left[\iint\limits_{G(u_1,u;\ 0,v_1)} a_j^{(u,v)}(s,t)\, a_k^{(u,v)}(s,t) \left(y^2(s,t) - \right. \right. \right.$$

$$\left. \left. \left. - E\, y^2(0,0) \right) ds\, dt \right]^2 \right\} \leq 2\, L_1^4 \left[r_1(0,0) + 2a^2 \right] c_{01}\, \frac{(u_2 - u_1)^2}{u_1^2\, v_1}, \tag{3.48}$$

Furthermore, by property 1a we have

$$E \left\{ \max_{\left((u,v) \in G(u_1,u_2;\ v_1,v_2) \right)} \times \right.$$

$$\times \left[\iint\limits_{G(u_1,u;\ v_1,v)} a_j^{(u,v)}(s,t)\, a_k^{(u,v)}(s,t)\, y^2(s,t)\, ds\, dt \right]^2 \right\} \leq$$

$$\leq \ E \left\{ \frac{L_1^4}{u_1^2 v_1^2} \left[\iint\limits_{G(u_1,u_2;\ v_1,v_2)} y^2(s,t)\, ds\, dt \right]^2 \right\} \leq$$

$$\leq \ L_1^4\, E\, y^4\,(0,0)\, \frac{(u_2 - u_1)^2\,(v_2 - v_1)^2}{u_1^2\, v_1^2}. \tag{3.49}$$

Now consider the inequality

$$\sup_{(u \geq S_0;\ v \geq T_0)} |\gamma_{jk}\,(u,v) - \gamma_{jk}| \leq \sup_{(\nu \geq \nu_0;\ \mu \geq \mu_0)} |\gamma_{jk}\,(\nu^2, \mu^2) - \gamma_{jk}| +$$

$$+ \sup_{(\nu \geq \nu_0; \, \mu \geq \mu_0)} \sup_{((u,v) \in G(\nu^2, (\nu+1)^2; \, \mu^2, (\mu+1)^2))} |\gamma_{jk}(u,v) - \gamma_{jk}(\nu^2, \mu^2)|,$$

(3.50)

where ν, μ, ν_0 and μ_0 are positive integers for which $\nu_0^2 \leq S_0 < (\nu_0 + 1)^2$ and $\mu_0^2 \leq T_0 < (\mu_0 + 1)^2$.

According to (3.46) and the Borel-Cantelli lemma we obtain

$$P \left\{ \lim_{\nu \to \infty, \, \mu \to \infty} \gamma_{jk}(\nu^2, \mu^2) = \gamma_{jk} \right\} = 1,$$

(3.51)

which makes it possible to bound the first term on the right-hand side of (3.50). To bound the second term of (3.50), consider the following representation:

$$\gamma_{jk}(u,v) - \gamma_{jk}(\nu^2, \mu^2) =$$

$$= \iint_{G(\nu^2, \, \mu^2)} a_j^{(u,v)}(s,t) \, a_k^{(u,v)}(s,t) \, y^2(s,t) \, ds \, dt \, -$$

$$- \iint_{G(\nu^2, \, \mu^2)} a_j^{(\nu^2, \mu^2)}(s,t) \, a_k^{(\nu^2, \mu^2)}(s,t) \, y^2(s,t) \, ds \, dt \, +$$

$$+ \iint_{G(\nu^2, u; \, 0, \mu^2)} a_j^{(u,v)}(s,t) \, a_k^{(u,v)}(s,t) \, [y^2(s,t) - E \, y^2(0,0)] \, ds \, dt \, +$$

$$+ \iint_{G(0, \nu^2; \, \mu^2, v)} a_j^{(u,v)}(s,t) \, a_k^{(u,v)}(s,t) \, [y^2(s,t) - E \, y^2(0,0)] \, ds \, dt \, +$$

$$+ \; E \{y^2(0,0)\} \iint_{G(\nu^2, u; \, 0, \mu^2) \bigcup G(0, \nu^2; \, \mu^2, v)} a_j^{(u,v)}(s,t) \, a_k^{(u,v)}(s,t) \, ds \, dt \, +$$

$$+ \iint\limits_{G(\nu^2,u;\,\mu^2,v)} a_j^{(u,v)}(s,t)\, a_k^{(u,v)}(s,t)\, y^2(s,t)\, ds\, dt. \tag{3.52}$$

Since for all $(u,v) \in G\left(\nu^2,\,(\nu+1)^2;\ \mu^2,\,(\mu+1)^2\right)$

$$\left| \iint\limits_{G(\nu^2,\mu^2)} a_j^{(u,v)}(s,t)\, a_k^{(u,v)}(s,t)\, y^2(s,t)\, ds\, dt- \right.$$

$$\left. - \iint\limits_{G(\nu^2,\mu^2)} a_j^{(\nu^2,\mu^2)}(s,t)\, a_k^{(\nu^2,\mu^2)}(s,t)\, y^2(s,t)\, ds\, dt \right| \le$$

$$\le \left[1 - \frac{q_j(\nu^2,\mu^2)\, q_k(\nu^2,\mu^2)}{q_j\left((\nu+1)^2,(\mu+1)^2\right) q_k\left((\nu+1)^2,(\mu+1)^2\right)} \right] \left|\gamma_{jk}(\nu^2,\mu^2)\right|$$

$$\tag{3.53}$$

and according to condition 1a the inequality

$$1 - \frac{q_j^2(\nu^2,\mu^2)}{q_j^2\left((\nu+1)^2,(\mu+1)^2\right)} =$$

$$= \iint\limits_{G((\nu+1)^2,(\mu+1)^2)-G(\nu^2,\mu^2)} \left[a_j^{((\nu+1)^2,(\mu+1)^2)}(s,t) \right]^2 ds\, dt \le$$

$$\le L_1^2\, \frac{(\nu+1)^2\,(\mu+1)^2 - \nu^2\,\mu^2}{(\nu+1)^2\,(\mu+1)^2},$$

holds, and, therefore,

$$\lim_{\nu\to\infty,\,\mu\to\infty} \frac{q_j\left(\nu^2,\mu^2\right)}{q_j\left((\nu+1)^2,(\mu+1)^2\right)} = 1, \quad 1 \le j \le m,$$

and taking into account (3.51) one can assert that the right side of (3.53) approaches 0 with probability 1 as ν, $\mu \to \infty$. Moreover, by 1a we have

$$
E\left\{y^2(0,0)\right\} \left| \iint\limits_{G(u_1,u;\,0,v_1)\bigcup G(0,u_1;\,v_1,v)} a_j^{(u,v)}(s,t)\, a_k^{(u,v)}(s,t)\, ds\, dt \right| \leq
$$

$$
\leq E\left\{y^2(0,0)\right\} L_1^2 \left[\frac{(u_2-u_1)\, v_1}{u_1\, v_1} + \frac{u_1\,(v_2-v_1)}{u_1\, v_1} \right] \tag{3.54}
$$

for $(u,v) \in G(u_1,u_2;\, v_1,v_2)$.

Therefore from (3.52) taking (3.47) - (3.49) and (3.54) into account and using the Borel-Cantelli lemma, we obtain

$$
\max_{(u,v)\in G\left(\nu^2,(\nu+1)^2;\;\mu^2,(\mu+1)^2\right)} \left|\gamma_{jk}(u,v) - \gamma_{jk}(\nu^2,\mu^2)\right| \to 0
$$

for ν, $\mu \to \infty$ with probability 1. This assertion along with (3.50) and (3.51) proves the lemma.

Now we shall describe the asymptotic properties of the estimator $\vec{\theta}\,(S,T)$.

Theorem 31 *Assume that conditions $1-3$ hold. Then the estimator $\vec{\theta}\,(S,T)$ is strongly consistent:*

$$
P\left\{ \lim_{S\to\infty,\,T\to\infty} \vec{\theta}\,(S,T) = \vec{\theta} \right\} = 1.
$$

Proof. Because

$$
\vec{\theta}\,(S,T) = \vec{\theta} + \Gamma^{-1}(S,T) \iint\limits_{G(S,T)} \widetilde{z}(s,t)\, \vec{a}\,(s,t)\, ds\, dt,
$$

where

$$
\widetilde{z}(s,t) = \left[\, x(s,t) - E\left\{x(s,t)/F\right\}\right] y(s,t),
$$

whence $E\,\widetilde{z}(s,t) = 0$, $(s,t) \in \mathcal{R}^2$, it is sufficient to prove the convergence to zero of the elements of the vector

$$\vec{\theta}\,(S,T) - \vec{\theta} = Q^{-\frac{1}{2}}(S,T)\,\Lambda(S,T)\,Q^{-\frac{1}{2}}(S,T) \times \iint\limits_{G(S,T)} \widetilde{z}(s,t)\,\vec{a}\,(s,t)\,ds\,dt$$

with probability 1. Here

$$\Lambda(S,T) = \left(\gamma^{jk}(S,T)\right)_{jk=1}^{m} = Q^{\frac{1}{2}}(S,T)\,\Gamma^{-1}(S,T)\,Q^{\frac{1}{2}}(S,T).$$

According to Lemma 18,

$$P\left\{\lim_{S\to\infty,\,T\to\infty} \gamma^{jk}(S,T) = \gamma^{jk}\right\} = 1,$$

where $(\gamma^{jk})_{jk=1}^{m} = \Gamma^{-1}$.

Furthermore, since the j-th element of the vector $\vec{\theta}\,(S,T) - \vec{\theta}$ has the form

$$\frac{1}{q_j(S,T)} \sum_{k=1}^{m} \gamma^{jk}(S,T) \iint\limits_{G(S,T)} \widetilde{z}(s,t)\,a_k^{(S,T)}(s,t)\,ds\,dt,$$

it is sufficient to verify for each $1 \le j, k \le m$ the relation

$$P\left\{\lim_{S\to\infty,\,T\to\infty} \frac{1}{q_j(S,T)} \iint\limits_{G(S,T)} \widetilde{z}(s,t)\,a_k^{(S,T)}(s,t)\,ds\,dt = 0\right\} = 1.$$

The proof of this relation can be done essentially the same way as that of Lemma 18, and we will omit it. We remark that an estimator of the form (3.46), for instance, can be obtained using properties of conditional expectation and the following consequence of property 1a: $q_k(u,v) \ge L^{(k)}\sqrt{uv}$ with some constant $L^{(k)} > 0$ for $u \ge u_0$ and $v \ge v_0$ with $u_0, v_0 > 0$.

Let

$$H = a^2 \iint\limits_{R^2} r(s,t)\,R(s,t)\,ds\,dt + \iint\limits_{R^2} r(s,t)\,r_1(s,t)\,R(s,t)\,ds\,dt.$$

Theorem 32 *Assume that conditions* $1 - 3$ *hold and* H *is positive definite.*
The distribution of the vector

$$Q^{\frac{1}{2}}(S,T) \left[\vec{\theta}\,(S,T) - \vec{\theta} \right]$$

converges weakly as $S, T \to \infty$ *to the normal distribution in* $\mathcal{R}^m$ *with mean* $\vec{0}$
and covariance matrix $F = \Gamma^{-1} H \Gamma^{-1}$.

Proof. Since

$$Q^{\frac{1}{2}}(S,T) \left[\vec{\theta}\,(S,T) - \vec{\theta} \right] = \Lambda(S,T)\,\vec{\eta}\,(S,T),$$

$$\vec{\eta}\,(S,T) = Q^{-\frac{1}{2}}(S,T) \iint\limits_{G(S,T)} \tilde{z}(s,t)\,\vec{a}\,(s,t)\,ds\,dt,$$

and by Lemma 18

$$P\left\{ \lim_{S \to \infty,\,T \to \infty} \Lambda(S,T) = \Gamma^{-1} \right\} = 1,$$

it is sufficient to consider only the behavior of the vector $\vec{\eta}\,(S,T)$ for $S, T \to \infty$.

Let $\vec{u}' = (u_1, \ldots, u_m) \in \mathcal{R}^m$ be a given vector. The random variable $\vec{u}'\,\vec{\eta}\,(S,T)$
posesses the following properties [31]: its conditional distribution with respect
to F is Gaussian; thus with probability 1

$$E\left\{ \vec{u}'\,\vec{\eta}\,(S,T)\,/\,F \right\} = 0,$$

$$E\left\{ [\vec{u}'\,\vec{\eta}\,(S,T)]^2\,/\,F \right\} = \iint\limits_{G(S,T)} \iint\limits_{G(S,T)} r(s - \tilde{s},\, t - \tilde{t})\,y(s,t)\,y(\tilde{s},\,\tilde{t})\,\vec{u}' \times$$

$$\times \vec{a}^{(S,T)}(s,t)\,\vec{a}^{(S,T)}(s,t)'\,\vec{u}\,ds\,dt\,d\tilde{s}\,d\tilde{t} = \Delta^2(S,T).$$

From this we have for the characteristic function φ_{ST} of the random variable
$\vec{u}'\,\vec{\eta}\,(S,T)$ for $\lambda \in \mathcal{R}$,

$$\varphi_{ST}(\lambda) = E \exp\left\{ i\lambda\,\vec{u}'\,\vec{\eta}\,(S,T) \right\} =$$

$$= E\left\{E\left[\exp\left\{i\lambda\,\vec{u}'\,\vec{\eta}\,(S,T)\,/\,F\right\}\right]\right\}=$$

$$= E\left\{\exp\left\{-\frac{\lambda^2}{2}\Delta^2(S,T)\right\}\right\}.$$

To prove the theorem it is sufficient to show that

$$\Delta^2(S,T)\to\vec{u}'\,H\,\vec{u},\quad S\to\infty,\quad T\to\infty$$

in probability, since from this assertion, on the basis of Lebesgue theorem on passing to the limit in (3.53), we obtain the desired conclusion in terms of characteristic functions.

The proof of (3.54) uses the formula of changing variables, conditions 1 and 2, Lebesgue limit theorem, and a simple computation, which we omit because it is cumbersome.

To prove the theorem of the existence of the moments of $\vec{\theta}\,(S,T)$ we first establish a series of auxiliary propositions. For a given vector $\vec{u}'=(u_1,\ldots,u_m)\in\mathcal{R}^m$ with $\parallel\vec{u}\parallel=1$ we put

$$h(s,t)=h(s,t;\,S,T)=\vec{u}'\,\vec{a}^{(S,T)}\,(s,t),$$

$$\Phi(\vec{u};\,S,T)=\iint\limits_{G(S,T)}y^2(u,v)\,h^2(u,v)\,du\,dv,$$

$$\psi(\tau;\,S,T)=E\left\{\exp\left[-\tau\,\Phi(\vec{u};\,S,T)\right]\right\},\quad\tau\geq0.$$

Lemma 19 *For any integer $l>0$ there exist $S_0,\,T_0>0$ such that, for $u>S_0$ and $v>T_0$, for some fixed number $\kappa>0$ the inequality*

$$\psi(\tau;\,u,v)\leq\left(1+\frac{\tau}{l}\right)^{-\kappa l},\quad\tau\geq0,$$

holds, in which the right-hand side does not depend on $\vec{u}$.

Proof. We use the general Karhunen-Loéve expansion. Let $\{\lambda_k^2,\,k\geq1\}$ be the sequence of eigenvalues and $\{\varphi_k,\,k\geq1\}$ the orthonormal sequence of eigenfunctions of the kernel $h(u,v)\,r_1(u-u_1,v-v_1)\,h(u_1,v_1)$, which is positive definite under our conditions when $(u, v, u_1, v_1) \in G(S,T) \times G(S,T)$ for fixed positive numbers S and T. Then for $(u, v) \in G(S,T)$ and $y_0(u,v) = y(u,v) - E\, y(u,v)$ we have

$$h(u,v)\, y_0(u,v) = \sum_{k=1}^{\infty} \lambda_k\, \xi_k\, \varphi_k(u,v); \quad (u,v) \in G(S,T), \qquad (3.55)$$

where the quantities ξ_k, $k \geq 1$, are defined by

$$\lambda_k\, \xi_k = \iint\limits_{G(S,T)} h(u,v)\, y_0(u,v)\, \varphi_k(u,v)\, du\, dv, \quad k \geq 1$$

and are independent Gaussian with $E\, \xi_k = 0$ and $E\, \xi_k^2 = 1$, $k \geq 1$. The series in (3.55) converges for each $(u, v) \in G(S,T)$ in mean square and with probability 1. The series in (3.55) may be multiplied by a function which is continuous on $G(S,T)$ and integrated term by term; the series obtained this way converges in mean square and with probability 1. It is also easy to verify that

$$\iint\limits_{G(S,T)} h^2(u,v)\, y_0^2(u,v)\, du\, dv = \sum_{k=1}^{\infty} \lambda_k^2\, \xi_k^2.$$

Putting

$$h_k = \iint\limits_{G(S,T)} h(u,v)\, \varphi_k(u,v)\, du\, dv, \quad k \geq 1,$$

we pass easily to

$$\Phi(\vec{u}; S,T) = a^2 \iint\limits_{G(S,T)} h^2(u,v)\, du\, dv + 2a \sum_{k=1}^{\infty} \lambda_k\, \xi_k\, h_k + \sum_{k=1}^{\infty} \lambda_k^2\, \xi_k^2.$$

Taking into account the properties of the random variables $\{\xi_k,\ k \geq 1\}$, with the aid of a simple calculation for the variables $\psi(\tau; S,T)$ we obtain

$$\psi(\tau; S,T) = \exp\left\{ -a^2\, \tau \iint\limits_{G(S,T)} h^2(u,v)\, du\, dv \right\} \times$$

$$\times \exp\left\{ \sum_{k=1}^{\infty} \frac{2a^2\,\tau\,\lambda_k^2\,h_k^2}{1+2\tau\,\lambda_k^2} \right\} \prod_{k=1}^{\infty} \left(1+2\tau\,\lambda_k^2\right)^{-\frac{1}{2}}.$$

From an obvious inequality and Bessel's inequality it follows that

$$a^2\,\tau \sum_{k=1}^{\infty} \frac{2\tau\,\lambda_k^2\,h_k^2}{1+2\tau\,\lambda_k^2} \le a^2\,\tau \sum_{k=1}^{\infty} h_k^2 \le a^2\,\tau \iint\limits_{G(S,T)} h^2(u,v)\,du\,dv;$$

thus

$$\psi(\tau;\,S,T) \le \prod_{k=1}^{\infty} \left(1+2\,\tau\,\lambda_k^2\right)^{-\frac{1}{2}}.$$

On the basis of the elementary inequality

$$(1+x\,y)^{1/\,lx} \ge 1 + \frac{y}{l}, \quad y \ge 0, \quad lx < 1, \quad x > 0$$

and the corollary of Mercer's theorem

$$L' = L'(S,T) = \sum_{k=1}^{\infty} \lambda_k^2 = r_1(0,0) \iint\limits_{G(S,T)} h^2(u,v)\,du\,dv$$

we have

$$\ln \psi\,(\tau;\,S,T) \le \ln \left(1+\frac{\tau}{l}\right)^{-lL'},$$

if only $2lL'' < 1$, where $L'' = L''(S,T) = \max\limits_{k\ge 1} \lambda_k^2$. By condition 1$b$ and the definition of h we have

$$\iint\limits_{G(S,T)} h^2(u,v)\,du\,dv \to \vec{u}'\, R(0,0)\, \vec{u}\ .$$

If $2k_1$ is the smallest eigenvalue of the positive definite (by condition 1b) matrix $R(0,0)$, then there exist $S_0, T_0 > 0$ such that for $S > S_0$ and $T > T_0$

$$\iint\limits_{G(S,T)} h^2(u,v)\,du\,dv > \kappa_1, \quad L' > r_1(0,0)\,\kappa_1 = \kappa > 0.$$

Making use of condition 2a, we easily establish that

$$\lambda_k^2 \le c_{11} \iint\limits_{G(S,T)} h^2(u,v)\, \varphi_k^2(u,v)\, du\, dv \le$$

$$\le c_{11} \max_{(u,v)\in G(S,T)} h^2(u,v) \le \frac{L_1^2 c_{11} \left(\sum\limits_{i=1}^{m}(u_i)\right)^2}{ST}, \quad k \ge 1.$$

Therefore there exist S_0 and T_0 such that $2lL'' < 1$ for $S > S_0$ and $T > T_0$. Lemma 19 is proved.

Using Lemma 19, we may also establish the following statement.

Lemma 20 *For any integer $l > 0$ there exist numbers S_0, T_0 such that, for $S > S_0$ and $T > T_0$ the inequality*

$$E\left\{\Phi^{-l}(\vec{u}; S, T)\right\} \le L$$

holds for some constant L which does not depend on $\vec{u}$, S, or T.

Let $\lambda(S,T)$ be the smallest eigenvalue of the matrix

$$Q^{-\frac{1}{2}}(S,T)\, \Gamma(S,T)\, Q^{-\frac{1}{2}}(S,T),$$

which is positive definite with probability 1; that is,

$$\lambda(S,T) = \min_{(\vec{u}:\|\vec{u}\|=1)} \Phi(\vec{u}; S, T).$$

Lemma 21 *For any integer $l > 0$ there are numbers S_0 and T_0 such that for $S > S_0$ and $T > T_0$ we have $E\left\{\lambda^{-1}(S,T)\right\} \le L$ for some constant L which does not depend on S and T.*

Proof. Let $A = \left\{\vec{u}: \|\vec{u}\| = 1\right\}$ and $U_\varepsilon(\vec{v}) = \left\{\vec{u}: \|\vec{u} - \vec{v}\| \le \varepsilon\right\}$ be the sphere of unit radius and the ε-neighborhood of the point $\vec{v}$ in $\mathcal{R}^m$. For any

$\varepsilon > 0$ let us consider the set of points $\left\{\vec{v}_1, \ldots, \vec{v}_n\right\}$ such that

$$\bigcup_{k=1}^{n} U_\varepsilon(\vec{v}_k) \supset A.$$

The number $n = n(\varepsilon)$ depends on ε; it is important that n may be chosen not too large. In fact an elementary computation shows that n may be chosen so that

$$n \leq c\varepsilon^{-m+1}, \quad c = \sqrt{\pi}\, m \cdot 3^{m-1}. \tag{3.56}$$

For a bound on $P\{\lambda(S,T) < z\}$ with $z > 0$ we have

$$P\left\{\lambda(S,T) < z\right\} = P\left\{\min_{(\|\vec{u}\|=1)} \Phi(\vec{u};\, S,T) < z\right\} \leq$$

$$\leq \sum_{j=1}^{n} P\left\{\min_{(\vec{u} \in \mathcal{U}_\varepsilon(\vec{v}_j) \cap A)} \Phi(\vec{u};\, S,T) < z\right\} \leq \sum_{j=1}^{n} \left[P\left\{\Phi(\vec{v}_j;\, S,T) < 2z\right\} + \right.$$

$$\left. + P\left\{\max_{(\vec{u} \in \mathcal{U}_\varepsilon(\vec{v}_j) \cap A)} \left[\Phi(\vec{u};\, S,T) - \Phi(\vec{v}_j;\, S,T) \right] \geq z\right\} \right]. \tag{3.57}$$

According to Lemma 20, for a given integer $\nu \geq 1$ there exist S_0 and T_0 such that for $S > S_0$ and $T > T_0$ we have

$$P\left\{\Phi(\vec{v}_j;\, S,T) < 2z\right\} \leq (2z)^\nu E\left\{\Phi^{-\nu}(\vec{v}_j;\, S,T)\right\} \leq Lz^\nu,$$

for $1 \leq j \leq n$ for a constant L which does not depend on S or T. Putting $\varepsilon = z^2$ and taking (3.56) into account, we get

$$\sum_{j=1}^{n} P\left\{\Phi(\vec{v}_j;\, S,T) < 2z\right\} \leq L_1 z^{\nu-2m+2}, \quad L_1 = cL. \tag{3.58}$$

Because the definition of $\Phi(\vec{u};\, S,T)$ implies for $1 \leq j \leq n$ the inequality

$$\left| \Phi(\vec{u};\, S,T) - \Phi(\vec{v}_j;\, S,T) \right| \leq 2\varepsilon \iint\limits_{G(S,T)} y^2(u,v) \, \| \vec{a}^{(S,T)}(u,v) \|^2 \, du\, dv,$$

from Čebyšev's inequality and condition $1a$ we get

$$P\left\{\max_{(\vec{u}\in\mathcal{U}_\epsilon(\vec{v}_j)\bigcap A)}\left[\Phi(\vec{u};S,T)-\Phi(\vec{v}_j;S,T)\right]\geq z\right\}\leq$$

$$\leq (2\epsilon)^\nu z^{-\nu} E\left\{\iint\limits_{G(S,T)} y^2(u,v)\,\|\,\vec{a}^{(S,T)}(u,v)\|^2\,du\,dv\right\}^\nu\leq$$

$$\leq (2\epsilon)^\nu z^{-\nu} L_1^{2\nu} E\{y^{2\nu}(0,0)\},$$

from which for $\epsilon = z^2$ we obtain

$$\sum_{j=1}^n P\left\{\max_{(\vec{u}\in\mathcal{U}_\epsilon(\vec{v}_j))\bigcap A}\left[\Phi(\vec{u};S,T)-\Phi(\vec{v}_j;S,T)\right]\geq z\right\}\leq L\,z^{\nu-2n+2}$$

(3.59)

for some constant L. The relations (3.57) -(3.59) allow us to conclude that for any integer $\nu \geq 1$ there are S_0 and T_0 such that for $S > S_0$ and $T > T_0$

$$P\{\lambda(S,T) < z\} \leq L\,z^{\nu-2m+2},\quad z > 0$$

for some constant L. From this it is evident that

$$P\{\lambda^{-1}(S,T) > z\} \leq L\,z^{-(\nu-2m+2)},\quad z > 0$$

whence Lemma 21 follows in a well-known way [28].

Theorem 33 *Assume that conditions $1-3$ hold and H is positive definite. Let $l \geq 1$ be a fixed number. There exist numbers $S(l)$ and $T(l)$ such that for $S > S(l)$ and $T > T(l)$, the moments of l-th order of the random vector*

$$Q^{-\frac{1}{2}}(S,T)\left[\vec{\theta}\,(S,T)-\vec{\theta}\right]$$

(3.60)

exist. In particular, for $S > S(l)$ and $T > T(l)$ the function $\vec{\theta}\,(S,T)$ is an unbiased estimator of $\vec{\theta}$. Moments of all orders of the variables (3.60) converge as $S,\,T \to \infty$ to the corresponding moments of a Gaussian distribution on $\mathcal{R}^m$ with mean $\vec{0}$ and covariance matrix F.

Proof. Let $\vec{u}' = (u_1, \ldots, u_m) \in \mathcal{R}^m$ be a fixed vector. The conditional distribution, given F, of the variable

$$\zeta(S,T) = \vec{u}' \, Q^{\frac{1}{2}}(S,T) \left[\vec{\theta}\,(S,T) - \vec{\theta} \right] =$$

$$= \vec{u}' \, Q^{\frac{1}{2}}(S,T) \, \Gamma^{-1}(S,T) \iint\limits_{G(S,T)} z(u,v) \, \vec{a}\,(u,v) \, du \, dv$$

is Gaussian [31], and

$$E\left\{\zeta(S,T)\,/\,F\right\} = 0,$$

$$E\left\{\zeta^2(S,T)\,/\,F\right\} = \vec{u}' \, Q^{\frac{1}{2}}(S,T) \, \Gamma^{-1}(S,T) \times$$

$$\times \iint\limits_{G(S,T)} r\,(u - \tilde{u},\, v - \tilde{v}) \, \vec{a}\,(u,v) \, \vec{a}\,(\tilde{u}, \tilde{v})' \, y(u,v)\, y(\tilde{u}, \tilde{v}) \times$$

$$\times \; du \, dv \, d\tilde{u} \, d\tilde{v} \, \Gamma^{-1}(S,T) \, Q^{\frac{1}{2}}(S,T) \, \vec{u}$$

with probability 1.

By condition 3 we have

$$E\left\{\zeta^2(S,T)\,/\,F\right\} \le c_0 \, \vec{u}' \, Q^{\frac{1}{2}}(S,T) \, \Gamma^{-1}(S,T) \, Q^{\frac{1}{2}}(S,T) \, \vec{u} \le c_0 \, \lambda\,(S,T)^{-1}.$$

Hence for $z > 0$ we get

$$P\left\{|\zeta(S,T)| \ge z\right\} \;=\; E\left\{P\left\{|\zeta(S,T)| \ge z/F\right\}\right\} \le$$

$$\le \; \frac{(2l)!}{2^l \, l! \, z^{2l}} \, E\left\{\left(E\left[\zeta^2(S,T)\,/\,F\right]\right)^l\right\} \le$$

$$\le \; \frac{(2l)! \, c_0^l}{2^l \, l! \, z^{2l}} \, E\left\{\lambda^{-l}(S,T)\right\}.$$

The latter inequality and Lemma 21 lead in a well-known way [100] to the existence and convergence of moments of estimator. The assertion of unbiasedness is now evident. Theorem is proved.

3.4 IDENTIFICATION OF THE PARAMETERS FOR THE STATIONARY NONLINEAR REGRESSION AS A SPECIAL CASE OF STOCHASTIC PROGRAMMING PROBLEM

We shall consider now nonlinear regression models for which Theorem 8 and Theorem 9 are applied.

Let us formulate some auxiliarly assertions.

Lemma 22 *[89] Let $\eta = \eta(\omega)$, $\zeta = \zeta(\omega)$ be real random variables on a probabilistic space $(\Omega, \mathcal{G}, P)$, and*

$$\eta(\omega) > \zeta(\omega), \quad \omega \in A, \quad A \in \mathcal{G}, \quad P(A) > 0.$$

Suppose that there exist

$$\int\limits_A \eta(\omega)\, dP, \quad \int\limits_A \zeta(\omega)\, dP.$$

Then

$$\int\limits_A \eta(\omega)\, dP > \int\limits_A \zeta(\omega)\, dP.$$

Lemma 23 *Let $\xi = \xi(\omega)$ be a real random variable on a probabilistic space $(\Omega, \mathcal{G}, P)$, and $E\,|\xi| < \infty$. Denote*

$$\varphi(c) = E\,|\xi - c|, \quad c \in \mathcal{R}.$$

Suppose that $a,\, b \in \mathcal{R}$. If a is a median of ξ, b is not a median of ξ then $\varphi(a) < \varphi(b)$; if a and b are medians of ξ then $\varphi(a) = \varphi(b)$.

Proof. Assume that $\alpha,\, \beta \in \mathcal{R}$, $\alpha < \beta$. Consider the probabilistic space $(\mathcal{R}, \mathcal{B}(\mathcal{R}), P_\xi)$, where

$$P_\xi(B) = P\,\{\xi \in B\}, \quad B \in \mathcal{B}(\mathcal{R}).$$

We have

$$\varphi(\beta) - \varphi(\alpha) = E\left(|\xi - \beta| - |\xi - \alpha|\right) = \int_{\mathcal{R}} \left(|x - \beta| - |x - \alpha|\right) dP_\xi =$$

$$= \int_{]-\infty,\alpha]} (\beta - \alpha)\, dP_\xi + \int_{]\alpha,\beta[} (\beta + \alpha - 2x)\, dP_\xi + \int_{[\beta,+\infty[} (\alpha - \beta)\, dP_\xi =$$

$$= (\beta - \alpha)\left(P_\xi\left(]-\infty,\alpha]\right) - P_\xi\left([\beta,+\infty[\right)\right) + \int_{]\alpha,\beta[} (\beta + \alpha - 2\,x)\, dP_\xi.$$

If $P_\xi\left(]\alpha, \beta[\right) = 0$ then

$$F_\xi(\beta) = P_\xi\left(]-\infty,\beta[\right) = P_\xi\left(]-\infty,\alpha]\right) = F_\xi(\alpha + 0);$$

$$\int_{]\alpha,\beta[} (\beta + \alpha - 2x)\, dP_\xi = 0.$$

Hence

$$\begin{aligned}
\varphi(\beta) - \varphi(\alpha) &= (\beta - \alpha)\left(F_\xi(\alpha + 0) - (1 - F_\xi(\beta))\right) = \\
&= (\beta - \alpha)\left(F_\xi(\alpha + 0) + F_\xi(\beta) - 1\right) = \\
&= (\beta - \alpha)\left(2\,F_\xi(\beta) - 1\right) = \\
&= (\beta - \alpha)\left(2\,F_\xi(\alpha + 0) - 1\right).
\end{aligned} \tag{3.61}$$

Let $P_\xi\left(]\alpha, \beta[\right) > 0$. For $x \in\,]\alpha, \beta[$

$$\alpha < x < \beta \Longrightarrow -2\,\alpha > -2\,x > -2\,\beta \Longrightarrow$$

$$\Longrightarrow \beta + \alpha - 2\,\alpha > \beta + \alpha - 2\,x > \beta + \alpha - 2\,\beta \Longrightarrow$$

$$\Longrightarrow \beta - \alpha > \beta + \alpha - 2\,x > \alpha - \beta.$$

By Lemma 22

$$\int_{]\alpha,\beta[} (\beta - \alpha)\, dP_\xi > \int_{]\alpha,\beta[} (\beta + \alpha - 2x)\, dP_\xi > \int_{]\alpha,\beta[} (\alpha - \beta)\, dP_\xi.$$

Thus,

$$(\beta - \alpha)\left(P_\xi\big(]-\infty,\,\alpha]\big) - P_\xi\big([\beta,\,+\infty[\big)\right) + (\beta - \alpha)\, P_\xi\big(]\alpha,\,\beta[\big) >$$

$$> \varphi(\beta) - \varphi(\alpha) > (\beta - \alpha)\left(P_\xi\big(]-\infty,\,\alpha]\big) - P_\xi\big([\beta,\,+\infty[\big)\right) +$$

$$+(\alpha - \beta)\, P_\xi\big(]\alpha,\,\beta[\big).$$

From the last relation we obtain

$$(\beta - \alpha)\left(P_\xi\big(]-\infty,\,\beta[\big) - P_\xi\big([\beta,\,+\infty[\big)\right) > \varphi(\beta) - \varphi(\alpha) >$$

$$> (\beta - \alpha)\left(P_\xi\big(]-\infty,\,\alpha]\big) - P_\xi\big(]\alpha,\,+\infty[\big)\right) \Longrightarrow$$

$$\Longrightarrow (\beta - \alpha)\left(F_\xi(\beta) - \big(1 - F_\xi(\beta)\big)\right) > \varphi(\beta) - \varphi(\alpha) >$$

$$> (\beta - \alpha)\left(F_\xi(\alpha + 0) - \big(1 - F_\xi(\alpha + 0)\big)\right) \Longrightarrow$$

$$\Longrightarrow (\beta - \alpha)\left(2F_\xi(\beta) - 1\right) >$$

$$> \varphi(\beta) - \varphi(\alpha) > (\beta - \alpha)\left(2F_\xi(\alpha + 0) - 1\right). \tag{3.62}$$

Consider for $\alpha, \beta \in \mathcal{R}$, $\alpha < \beta$ three variants:

1) α is a median of ξ, β is not a median of ξ;

2) α is not a median of ξ, β is a median of ξ;

3) α and β are medians of ξ.

Suppose that α is a median of ξ, β is not a median of ξ. Then $F_\xi(\beta) > 1/2$. If $P_\xi\left(\,]\alpha,\,\beta[\,\right) = 0$ then by (3.61)

$$\varphi(\beta) - \varphi(\alpha) = (\beta - \alpha)\left(2\,F_\xi\,(\beta) - 1\right) > 0.$$

For $P_\xi\left(\,]\alpha,\,\beta[\,\right) > 0$ by (3.62) we have

$$\varphi(\beta) - \varphi(\alpha) > (\beta - \alpha)\left(2\,F_\xi\,(\alpha + 0) - 1\right) \geq 0.$$

Then $\varphi(\alpha) < \varphi(\beta)$.

Assume that α is not a median of ξ, β is a median. In this case $F_\xi\,(\alpha + 0) < 1/2$. By (3.61), (3.62)

$$\varphi(\beta) - \varphi(\alpha) = (\beta - \alpha)\left(2\,F_\xi\,(\alpha + 0) - 1\right) < 0, \quad P_\xi\left(\,]\alpha,\,\beta[\,\right) = 0;$$

$$\varphi(\beta) - \varphi(\alpha) < (\beta - \alpha)\left(2\,F_\xi\,(\beta) - 1\right) \leq 0, \quad P_\xi\left(\,]\alpha,\,\beta[\,\right) > 0.$$

Hence $\varphi(\beta) < \varphi(\alpha)$.

If α and β are medians then

$$1/2 \leq F_\xi\,(\alpha + 0) \leq F_\xi(\beta) \leq 1/2.$$

Thus,

$$F_\xi\,(\alpha + 0) = F_\xi(\beta) = 1/2.$$

We obtain

$$P_\xi\left(\,]\alpha,\,\beta[\,\right) = F_\xi(\beta) - F_\xi\,(\alpha + 0) = 0 \Longrightarrow$$

$$\Longrightarrow \varphi(\beta) - \varphi(\alpha) = (\beta - \alpha)\left(2\,F_\xi\,(\beta) - 1\right) = 0.$$

Consequently,

$$\varphi(\alpha) = \varphi(\beta).$$

The lemma is proved.

Suppose that $(\vec{x}_i, \vec{y}_i)$, $i \geq 1$ are independent identically distributed random vectors on a complete probabilistic space $(\Omega, \mathcal{G}, P)$ with values in $\mathcal{R}^{k+m}$; $\vec{x}_i \in \mathcal{R}^k$, $\vec{y}_i \in \mathcal{R}^m$; $k, m \geq 1$.

Denote

$$\|\vec{a}\|_1 = \sum_{j=1}^{k} |a_j|, \quad \vec{a} = (a_j)_{j=1}^{k} \in \mathcal{R}^k.$$

We will use the following assumptions.

1. For every $j \in \{\overline{1, k}\}$

$$P\left\{x_{j1} < f_j(\vec{\theta}, \vec{y}_1) / \vec{y}_1\right\} \leq 1/2;$$

$$P\left\{x_{j1} \leq f_j(\vec{\theta}, \vec{y}_1) / \vec{y}_1\right\} \geq 1/2$$

with probability 1, where $\vec{x}_1 = (x_{j1})_{j=1}^{k}$; θ is a vector from I, I is some closed subset of $\mathcal{R}^l$, $l \geq 1$; $\vec{f} = (f_j)_{j=1}^{k} : I \times \mathcal{R}^m \to \mathcal{R}^k$ is some known function, which is continuous in the first argument and measurable in the second argument.

2. $E \left\|\vec{x}_1\right\|_1 < \infty.$

3. For all $c > 0$

$$E\left\{\max_{\|\vec{u}\| \leq c} \left\|\vec{f}(\vec{u}, \vec{y}_1)\right\|_1\right\} < \infty.$$

4. If I is unbounded then for any $\vec{z} \in A$, $P\left\{\vec{y}_1 \in A\right\} = 1$

$$\left\|\vec{f}(\vec{u}, \vec{z})\right\|_1 \to \infty, \quad \|\vec{u}\| \to \infty.$$

Denote

$$\vec{\eta} = (\eta_j)_{j=1}^{k} = \vec{x}_1 - \vec{f}(\vec{\theta}, \vec{y}_1).$$

Let $q_j(B, \vec{z})$, $B \in \mathcal{B}(\mathcal{R})$, $\vec{z} \in \mathcal{R}^m$ be the regular conditional distribution of the random variable η_j with condition $\vec{y}_1 = \vec{z}$ [31], $j = \overline{1, k}$.

5. For every $\vec{u} \in I$, $\vec{u} \neq \vec{\theta}$ there exist an index $j_0 = j_0(\vec{u}) \in \{\overline{1,k}\}$ and a set $C = C(\vec{u}) \subset \mathcal{R}^m$, $P\left\{\vec{y}_1 \in C\right\} > 0$ such that for all $\vec{z} \in C$ the value $\beta = f_{j_0}(\vec{u}, \vec{z}) - f_{j_0}(\vec{\theta}, \vec{z})$ is not a median of the probability distribution $q_{j_0}(B, \vec{z})$, $B \in \mathcal{B}(\mathcal{R})$.

We need to estimate $\vec{\theta}$ by the observations

$$\left\{(\vec{x}_i, \vec{y}_i), \quad i = \overline{1,n}\right\}, \quad n \in \mathcal{N}.$$

Consider the least modules estimate which is a minimum point of the function

$$F_n(\vec{u}) = \frac{1}{n} \sum_{i=1}^{n} \left\| \vec{x}_i - \vec{f}(\vec{u}, \vec{y}_i) \right\|_1, \quad \vec{u} \in I. \tag{3.63}$$

Theorem 34 *If assumptions $1 - 5$ hold then for all $n \in \mathcal{N}$ and $\omega \in \Omega'$, $P(\Omega') = 1$ there exists at least one minimum point $\vec{\theta}_n = \vec{\theta}_n(\omega)$ of the function (3.63), and for each n the mapping $\vec{\theta}_n(\omega)$, $\omega \in \Omega'$ can be chosen as $\mathcal{G}'_n$-measurable, where*

$$\mathcal{G}'_{\vec{n}} = \mathcal{G}_{\vec{n}} \bigcap \Omega', \quad \mathcal{G}_{\vec{n}} = \sigma\left\{ (\vec{x}_i, \vec{y}_i), i = \overline{1,n} \right\}.$$

For an arbitrary choice of the $\mathcal{G}'_n$-measurable function $\vec{\theta}_n(\omega)$

$$P\left\{ \vec{\theta}_n \to \vec{\theta}, \quad F_n(\vec{\theta}_n) \to F(\vec{\theta}), \quad n \to \infty \right\} = 1,$$

where $F(\vec{u}) = E\left\| \vec{x}_1 - \vec{f}(\vec{u}, \vec{y}_1) \right\|_1$, $\vec{u} \in I$.

Proof. We will show that $\vec{\theta}$ is a single point of minimum for function $F(\vec{u})$, $\vec{u} \in I$.

Fix $\vec{u} \in I$, $\vec{u} \neq \vec{\theta}$. We have

$$F(\vec{u}) - F(\vec{\theta}) = E\left\| \vec{\eta} - \left(\vec{f}(\vec{u}, \vec{y}_1) - \vec{f}(\vec{\theta}, \vec{y}_1) \right) \right\|_1 - E\|\vec{\eta}\|_1 = \sum_{j=1}^{k} E\Delta_j, \tag{3.64}$$

where

$$\Delta_j = E\left(\left|\eta_j - \left(f_j(\vec{u}, \vec{y}_1) - f_j(\vec{\theta}, \vec{y}_1)\right)\right| / \vec{y}_1\right) - E\left(|\eta_j| / \vec{y}_1\right), \quad j = \overline{1, k}.$$

By the properties of regular conditional distributions [31]

$$\Delta_j = \int_{\mathcal{R}} \left|s - \left(f_j(\vec{u}, \vec{y}_1) - f_j(\vec{\theta}, \vec{y}_1)\right)\right| q_j(ds, \vec{y}_1) - \int_{\mathcal{R}} |s| q_j(ds, \vec{y}_1),$$

$$j = \overline{1, k}$$

with probability 1.

By virtue of assumption 1 zero is a median of the distributions $q_j(B, \vec{y}_1)$, $B \in \mathcal{B}(\mathcal{R})$, $j = \overline{1, k}$ almost certainly. Then Lemma 23 implies that $\Delta_j \geq 0$, $j = \overline{1, k}$ with probability 1. It follows from assumption 5 and Lemma 23 that $P\{\Delta_{j_0} > 0\} > 0$. We obtain

$$E\,\Delta_j \geq 0, \quad j = \overline{1, k}; \quad E\,\Delta_{j_0} > 0.$$

Then from (3.64) we have $F(\vec{u}) > F(\vec{\theta})$.

Now the theorem follows from Theorem 8. Clearly, for random vectors $\vec{\xi}_i = (\vec{x}_i, \vec{y}_i)$, $i \geq 1$ and the function

$$f(\vec{u}, \vec{z}) = \left\|\vec{z}_1 - \vec{f}(\vec{u}, \vec{z}_2)\right\|_1, \quad \vec{u} \in I \subset \mathcal{R}^l, \quad \vec{z} = (\vec{z}_1, \vec{z}_2) \in \mathcal{R}^{k+m}$$

all conditions of Theorem 8 are fulfilled. The proof is complete.

Let us suppose now that for independent identically distributed random vectors $(\vec{x}_i, \vec{y}_i)$, $i \geq 1$ with values in $\mathcal{R}^{k+m}$ the following assumptions are valid.

1. $E\left\{x_{j1} / \vec{y}_1\right\} = f_j(\vec{\theta}, \vec{y}_1), j = \overline{1, k}$

   with probability 1, where $\vec{x}_1 = (x_{j1})_{j=1}^k$, $\vec{\theta} \in I$, I is a closed subset of $\mathcal{R}^l$, $\vec{f} = (f_j)_{j=1}^k \ : \ I \times \mathcal{R}^m \rightarrow \mathcal{R}^k$ is continuous in the first argument and measurable in the second argument.

2. $E\|\vec{x}_1\|^2 < \infty$.

3. For every $c > 0$

$$E \left\{ \max_{\|\vec{u}\| \leq c} \left\| \vec{f}(\vec{u}, \vec{y}_1) \right\|^2 \right\} < \infty.$$

4. If I is unbounded then for all $\vec{z} \in A$, $P\left\{ \vec{y}_1 \in A \right\} = 1$ we have

$$\left\| \vec{f}(\vec{u}, \vec{z}) \right\| \to \infty, \quad \|\vec{u}\| \to \infty.$$

5. For any $\vec{u} \in I$, $\vec{u} \neq \vec{\theta}$

$$E \left\{ \left\| \vec{f}(\vec{u}, \vec{y}_1) - \vec{f}(\vec{\theta}, \vec{y}_1) \right\|^2 \right\} > 0.$$

We will consider the least squares estimate of $\vec{\theta}$ by the observations $\left\{ (\vec{x}_i, \vec{y}_i), \quad i = \overline{1, n} \right.$
It is a minimum point of the function

$$F_n(\vec{u}) = \frac{1}{n} \sum_{i=1}^{n} \left\| \vec{x}_i - \vec{f}(\vec{u}, \vec{y}_i) \right\|^2, \quad \vec{u} \in I. \tag{3.65}$$

Theorem 35 *For any $n \in \mathcal{N}$, $\omega \in \Omega'$, $P(\Omega') = 1$ there exists at least one minimum point $\vec{\theta}_n = \vec{\theta}_n(\omega)$ of the function (3.65), and for every n the mapping $\vec{\theta}_n$ (ω) can be chosen $\mathcal{G}'_n$-measurable, $\mathcal{G}'_n = \mathcal{G}_n \cap \Omega'$, $\mathcal{G}_n = \sigma \left\{ (\vec{x}_i, \vec{y}_i), \quad i = \overline{1, n} \right\}$. Then*

$$P\left\{ \vec{\theta}_n \to \vec{\theta}, \quad F_n(\vec{\theta}_n) \to F(\vec{\theta}), \quad n \to \infty \right\} = 1,$$

where $F(\vec{u}) = E \left\| \vec{x}_1 - \vec{f}(\vec{u}, \vec{y}_1) \right\|^2$.

Proof. Let $\vec{u} \neq \vec{\theta}$. Then by the properties of mathematical expectations and condition 5

$$F(\vec{u}) - F(\vec{\theta}) = E \left\| \vec{x}_1 - \vec{f}(\vec{u}, \vec{y}_1) \right\|^2 - E \left\| \vec{x}_1 - \vec{f}(\vec{\theta}, \vec{y}_1) \right\|^2 =$$

$$= E\left(\left\|\vec{x}_1 - \vec{f}\left(\vec{u}, \vec{y}_1\right)\right\|^2 - \left\|\vec{x}_1 - \vec{f}\left(\vec{\theta}, \vec{y}_1\right)\right\|^2\right) =$$

$$= E\left(\sum_{j=1}^{k}\left(x_{j1} - f_j(\vec{u}, \vec{y}_1)\right)^2 - \sum_{j=1}^{k}\left(x_{j1} - f_j(\vec{\theta}, \vec{y}_1)\right)^2\right) =$$

$$= E\sum_{j=1}^{k}\left(\left(x_{j1} - f_j(\vec{u}, \vec{y}_1)\right)^2 - \left(x_{j1} - f_j(\vec{\theta}, \vec{y}_1)\right)^2\right) =$$

$$= \sum_{j=1}^{k} E\left(-2x_{j1}f_j(\vec{u}, \vec{y}_1) + f_j^2(\vec{u}, \vec{y}_1) + 2x_{j1}f_j(\vec{\theta}, \vec{y}_1) - f_j^2(\vec{\theta}, \vec{y}_1)\right) =$$

$$= \sum_{j=1}^{k} E\left(-2f_j(\vec{\theta}, \vec{y}_1)f_j(\vec{u}, \vec{y}_1) + f_j^2(\vec{u}, \vec{y}_1) + 2f_j^2(\vec{\theta}, \vec{y}_1) - f_j^2(\vec{\theta}, \vec{y}_1)\right) =$$

$$= \sum_{j=1}^{k} E\left(f_j(\vec{u}, \vec{y}_1) - f_j(\vec{\theta}, \vec{y}_1)\right)^2 > 0.$$

Denote

$$\vec{\xi}_i = (\vec{x}_i, \vec{y}_i); \quad f(\vec{u}, \vec{z}) = \left\|\vec{z}_1 - \vec{f}\left(\vec{u}, \vec{z}_2\right)\right\|^2, \quad \vec{z} = (\vec{z}_1, \vec{z}_2).$$

Now the theorem follows from Theorem 8.

Theorem 36 *Let assumptions 1–5 and the following conditions be valid:*

1) $\vec{\theta}$ *is an internal point of I;*

2) *there exists such a closed neighbourhood S of $\vec{\theta}$ that for any $\vec{z} \in \mathcal{R}^m$ the function $\vec{f}(\vec{u}, \vec{z})$ is twice continuously differentiable on S;*

3) $E\left\{\max\limits_{\vec{u}\in S}\left(\dfrac{\partial f_p}{\partial u_j}(\vec{u},\vec{y}_1)\right)^2\right\}<\infty,\quad j=\overline{1,l},\quad p=\overline{1,k},$

where $\vec{u}=(u_j)^l_{j=1};$

4) $E\left\{\max\limits_{\vec{u}\in S}\left(\dfrac{\partial^2 f_p}{\partial u_j\,\partial u_q}(\vec{u},\vec{y}_1)\right)^2\right\}<\infty,\quad j,q=\overline{1,l},\quad p=\overline{1,k};$

5) $E\left\{\parallel\vec{x}_1\parallel^4\right\}<\infty;$

6) $E\left\{\parallel\vec{f}(\vec{\theta},\vec{y}_1)\parallel^4\right\}<\infty;$

7) $E\left\{\left(\dfrac{\partial f_p}{\partial u_j}(\vec{\theta},\vec{y}_1)\right)^4\right\}<\infty,\quad j=\overline{1,l},\quad p=\overline{1,k};$

8) $\det A\neq 0,$ where

$$A=\left(\sum_{p=1}^{k}E\left(\dfrac{\partial f_p}{\partial u_i}(\vec{\theta},\vec{y}_1)\,\dfrac{\partial f_p}{\partial u_j}(\vec{\theta},\vec{y}_1)\right)\right)^l_{i,j=1};$$

9) $\det C\neq 0,$ where

$$C=E\left(\left(\dfrac{\partial f_p}{\partial u_i}(\vec{\theta},\vec{y}_1)\right)^{i=\overline{1,l}}_{p=\overline{1,k}}\left(\vec{x}_1-\vec{f}(\vec{\theta},\vec{y}_1)\right)\times\right.$$

$$\left.\times\left(\vec{x}_1-\vec{f}(\vec{\theta},\vec{y}_1)\right)'\left(\dfrac{\partial f_p}{\partial u_i}(\vec{\theta},\vec{y}_1)\right)^{p=\overline{1,k}}_{i=\overline{1,l}}\right).$$

Then $\sqrt{n}(\vec{\theta}_n-\vec{\theta})$ converges weakly to $N\left(\vec{0},A^{-1}CA^{-1}\right),\ n\to\infty.$

Suppose that the following condition also takes place:

10) $\sigma^2=E\left\{\left(\parallel\vec{x}_1-\vec{f}(\vec{\theta},\vec{y}_1)\parallel^2-F(\vec{\theta})\right)^2\right\}>0.$

Then $\sqrt{n}\left(F_n(\vec{\theta}_n)-F(\vec{\theta})\right)$ converges weakly to $N\left(0,\sigma^2\right),\ n\to\infty.$

The theorem follows from Theorem 9.

Analogs of Theorems 34, 35 for the case when $(\overrightarrow{x}_i, \overrightarrow{y}_i)$ is an ergodic stationary in a strict sense random process with discrete time are formulated easily. Similar results take place for measurable stationary in a strict sense ergodic random process $\left\{ \left(\overrightarrow{x}(t), \overrightarrow{y}(t) \right), \ t \in \mathcal{R} \right\}$. In this case

$$F_T(\overrightarrow{u}) = \frac{1}{T} \int\limits_0^T \left\| \overrightarrow{x}(t) - \overrightarrow{f}\left(\overrightarrow{u}, \overrightarrow{y}(t) \right) \right\|_1 dt \to \min$$

or

$$F_T(\overrightarrow{u}) = \frac{1}{T} \int\limits_0^T \left\| \overrightarrow{x}(t) - \overrightarrow{f}\left(\overrightarrow{u}, \overrightarrow{y}(t) \right) \right\|^2 dt \to \min.$$

Analog of Theorem 36 for ergodic stationary random processes can be formulated too, but it requires additional investigations, and we will not consider it.

It is worth to note that analogous results for the least squares estimates are contained in [15].

3.5 NONSTATIONARY REGRESSION MODEL FOR A RANDOM FIELD OBSERVED IN A CIRCLE

We will investigate properties of the least squares estimates for unknown parameters of regression for a homogeneous random field by its observations in the circle $s^2 + t^2 \le r^2$. Suppose that the observations model has the following form:

$$y(s, t) = \left[\sum_{k=1}^m \alpha_k \varphi_k(s, t) \right] \xi(s, t) + \eta(s, t) = \zeta(s, t) + \eta(s, t).$$

For the case $\xi(t) \equiv 1$ properties of the estimates are investigated for a discrete and continuous case. Asymptotic properties of the estimates for this case are considered in [36], [42], [44]. The cases when $(\xi(t), \eta(t))$ are Gaussian are investigated in [15]. We do not suppose that the field $\eta(s,t)$ is Gaussian, we assume that it satisfies the strong mixing condition.

We need the following assumptions.

1. Let

- $\vec{\alpha}' = (\alpha_1, \ldots, \alpha_m)$ be a vector of real numbers,

- $(\varphi_1(s,t), \ldots, \varphi_m(s,t))$ be a fixed vector of real continuous functions in $\mathcal{R}^2$,

- $\vec{\varphi}' = (\varphi_1, \ldots, \varphi_m)$.

Denote

- $q_k^2(r) = \iint\limits_{s^2 + t^2 \leq r^2} \varphi_k^2(s,t) \, ds \, dt,$

- $G(r_1, r_2) = \left\{ (s,t) : r_1 \leq s^2 + t^2 \leq r_2 \right\},$

- $G(r) = \left\{ (s,t) : s^2 + t^2 \leq r^2 \right\},$

- $Q(r) = \left(q_k^2(r) \delta_{jk} \right)_{j,k=1}^m,$

- $\vec{\varphi^r}(s,t) = Q^{-1/2}(r) \, \vec{\varphi}(s,t).$

1a. For some $L_1 > 0$ and all $1 \leq k \leq m$ and $r > 0$

$$q_k^{-1}(r) \max_{(s,t) \in G(r)} \varphi_k(s,t) \leq \frac{L_1}{r}.$$

1b. For some $L_2 > 0$ and any s, t and $r > 0$

$$\left\| \iint\limits_{G(r)} \vec{\varphi^r}(s_1 + |s|, t_1 + |t|) \, \vec{\varphi^r}{}'(s_1, t_1) \, ds_1 \, dt_1 - B(s,t) \right\| \leq \frac{L_2}{r}$$

with some matrix $B(s,t) = (B_{jk}(s,t))_{j,k=1}^m$, elements of which are continuous functions of (s,t), and $B(0,0)$ is positive definite.

1c. For some $L_3 > 0$ and any $r > 0$

$$\frac{q_k^2(2r)}{q_k^2(r)} \leq L_3 r, \quad k = \overline{1, m}.$$

1d. The functions $\varphi_k(s, t)$ are linearly independent on sets of positive Lebesgue measure.

2. The random fields $\xi(s, t)$ and $\eta(s, t)$ are independent homogeneous and isotropic in a strict sense,

$$E\,\xi(s, t) = a, \quad R(r) = E\left[\xi(s, t) - a\right]\left[\xi(0, 0) - a\right], \quad E\,\eta(s, t) = 0,$$

$$R_1(r) = E\,\eta(s, t)\,\eta(0, 0), \quad r^2 = s^2 + t^2,$$

and the field $\eta(s, t)$ satisfies the strong mixing condition with a coefficient
$$\psi(d) \leq \frac{C}{d^{2+\varepsilon}}, \quad \varepsilon > 0.$$

3. $E\,|\,\eta(s, t)\,|^{2+\delta} < \infty, \quad \delta > \dfrac{8}{\varepsilon}.$

4. The random field $\xi(s, t)$ is Gaussian and

$$\int\limits_0^\infty r\,R(r)\,dr \leq c < \infty.$$

5. The function $R(r)$ satisfies the following positive definition condition: for any $r_1, r_2 > 0$ a function $g(s, t) \in L_2\left(G(r)\right)$ with

$$\int\limits_{G(r)}\int\limits_{G(r)} R\left(\sqrt{(s_1 - s_2)^2 + (t_1 - t_2)^2}\right) g(s_1, t_1)\,g(s_2, t_2)\,ds_1\,dt_1\,ds_2\,dt_2 = 0$$

is equal to zero almost everywhere on $G(r)$ in the sense of Lebesgue measure in $\mathcal{R}^2$.

As an estimate of the unknown vector $\vec{\alpha'}$ by observations $\left\{\xi(s, t), y(s, t), (s, t) \in G(r)\right\}$ we consider the least squares estimate

$$\vec{\alpha}\,(r) = \Gamma^{-1}(r)\iint\limits_{G(r)} y(u, v)\,\xi(u, v)\,\vec{\varphi}\,(u, v)\,du\,dv,$$

where

$$\Gamma(r) = \iint\limits_{G(r)} \vec{\varphi}\,(u,v) \left(\vec{\varphi}\,(u,v)\right)' \xi(u,v)\,du\,dv.$$

By conditions 1d and 5 for $r > 0$ the matrix $\Gamma(r)$ is positively defined with probability 1. Thus, the matrix $\Gamma^{-1}(r)$ and, consequently, the estimate $\vec{\alpha}\,(r)$ are defined with probability 1.

The following assertion is valid.

Lemma 24 *Let conditions* 1a, 1b, 4 *and* 5 *be fulfilled. Then*

$$P\left\{\lim_{r\to\infty} Q^{-1/2}(r)\Gamma(r)\,Q^{-1/2}(r) = \Gamma\right\} = 1$$

with the positive definite matrix

$$\Gamma = (\gamma_{jk})_{j,k=1}^m = \left[a^2 + R(0)\right] B(0,0).$$

Proof. Let

$$\gamma_{jk}(r) = \iint\limits_{G(r)} \varphi_j^r(s,t)\,\varphi_k^r(s,t)\,\xi^2(s,t)\,ds\,dt.$$

By conditions 1a, 1b and 4

$$E\left|\gamma_{jk}(r) - \gamma_{jk}\right|^2 \leq$$

$$\leq\; 2\left[E\,\gamma_{jk}(r) - \gamma_{jk}\right]^2 + 2E\left|\gamma_{jk}(r) - E\,\gamma_{jk}(r)\right| =$$

$$=\; 2\left[E\,\xi^2(0,0)\right]^2 \left[\iint\limits_{G(r)} \varphi_j^r(s,t)\,\varphi_k^r(s,t)\,ds\,dt - B_{jk}\right]^2 +$$

$$+ E \left\{ \iint\limits_{G(r)} \varphi_j^r(s,t)\, \varphi_k^r(s,t)\, \left[\xi^2(s,t) - R(0) - a^2\right] ds\, dt \right\}^2 =$$

$$= 2\left[E\xi^2(0,0)\right]^2 \left[\iint\limits_{G(r)} \varphi_j^r(s,t)\, \varphi_k^r(s,t)\, ds\, dt - B_{jk} \right]^2 +$$

$$+ 4 \iint\limits_{G(r)} \iint\limits_{G(r)} \varphi_j^r(s,t)\, \varphi_k^r(s,t)\, \varphi_j^r(\widetilde{s},\widetilde{t})\, \varphi_k^r(\widetilde{s},\widetilde{t}) \times$$

$$\times \left[R^2 \left(\sqrt{(s-\widetilde{s})^2 + (t-\widetilde{t})^2} \right) + \right.$$

$$\left. + 2a^2 R \left(\sqrt{(s-\widetilde{s})^2 + (t-\widetilde{t})^2} \right) \right] ds\, dt\, d\widetilde{s}\, d\widetilde{t} \le$$

$$\le \frac{L'}{r^2} + \frac{L''}{r^4} \iint\limits_{G(r)} \iint\limits_{G(r)} \left[R^2 \left(\sqrt{(s-\widetilde{s})^2 + (t-\widetilde{t})^2} \right) + \right.$$

$$\left. + 2a^2 R \left(\sqrt{(s-\widetilde{s})^2 + (t-\widetilde{t})^2} \right) \right] ds\, dt\, d\widetilde{s}\, d\widetilde{t} \le \frac{L}{r^2}.$$

Then

$$E \left\{ \max_{r_1 \le r \le r_2} \left[\iint\limits_{G(r_1,r)} \varphi_j^r(s,t)\, \varphi_k^r(s,t)\, \xi^2(s,t)\, ds\, dt \right]^2 \right\} \le$$

$$\le E \left\{ \max_{r_1 \le r \le r_2} \frac{1}{r^4} \left[\iint\limits_{G(r_1,r)} \xi^2(s,t)\, ds\, dt \right]^2 \right\} \le$$

$$\leq \; LE\,\xi^4(0,0)\,\pi^2\,(r_2^2 - r_1^2)^2\,r_1^{-4}. \tag{3.66}$$

It follows from (3.66) that

$$P\left\{\lim_{n\to\infty}\gamma_{jk}(n) = \gamma_{jk}\right\} = 1. \tag{3.67}$$

Because of

$$\frac{[(n+1)^2 - n^2]^2}{n^4} = \frac{(2n+1)^2}{n^4}$$

one has

$$P\left\{\lim_{n\to\infty}\sup_{n\leq r\leq n+1}\left|\iint\limits_{G(n,r)} \varphi_j^r(s,t)\,\varphi_k^r(s,t)\,\xi^2(s,t)\,ds\,dt\right|^2 = 0\right\} = 1.$$

The following inequality is valid for $n < r < n+1$:

$$\left|\gamma_{jk}(r) - \gamma_{jk}(n)\right| = \left|\iint\limits_{G(r)}\varphi_j^r(s,t)\,\varphi_k^r(s,t)\,\xi^2(s,t)\,ds\,dt - \right.$$

$$\left. - \iint\limits_{G(n)}\varphi_j^n(s,t)\,\varphi_k^n(s,t)\,\xi^2(s,t)\,ds\,dt\right| \leq \left|\iint\limits_{G(n,r)}\varphi_j^r(s,t)\,\varphi_k^r(s,t)\,\xi^2(s,t)\,ds\,dt\right| +$$

$$+ \left[\frac{1}{q_j(n)\,q_k(n)} - \frac{1}{q_j(r)\,q_k(r)}\right]\left|\iint\limits_{G(r)}\varphi_j(s,t)\,\varphi_k(s,t)\,\xi^2(s,t)\,ds\,dt\right| \leq$$

$$\leq \left|\iint\limits_{G(r)}\varphi_j^r(s,t)\,\varphi_k^r(s,t)\,\xi^2(s,t)\,ds\,dt\right| + \frac{q_j(n+1)\,q_k(n+1) - q_j(n)\,q_k(n)}{q_j(n)\,q_k(n)} \times$$

$$\times \left[|\gamma_j| + |\gamma_{jk}(r) - \gamma_{jk}|\right]. \tag{3.68}$$

By condition 1a

$$1 - \frac{q_j^2(n)}{q_j^2(n+1)} = \frac{q_j^2(n+1) - q_j^2(n)}{q_j^2(n+1)} =$$

$$= \frac{1}{q_j^2(n+1)} \iint\limits_{G(n,n+1)} \varphi_j^2(s,t)\,ds\,dt \leq$$

$$\leq \frac{L}{(n+1)^2}\left[(n+1)^2 - n^2\right] \to 0$$

as $n \to \infty$. Consequently,

$$\lim_{n\to\infty} \frac{q_j(n)}{q_j(n+1)} = 1.$$

Then

$$\sup_{r \geq r_0}\left|\gamma_{jk}(r) - \gamma_{jk}\right| \leq \sup_{n \geq r_0}\left|\gamma_{jk}(n) - \gamma_{jk}\right| + \sup_{n \geq r_0} \sup_{n \leq r \leq n+1}\left|\gamma_{jk}(r) - \gamma_{jk}(n)\right|. \quad (3.69)$$

From (3.66) - (3.69) we obtain the assertion of the lemma.

Theorem 37 *Let conditions 1 - 5 be fulfilled. Then*

$$P\left\{\lim_{r\to\infty} \vec{\alpha}\,(r) = \vec{\alpha}\right\} = 1.$$

Proof. The following representation is valid:

$$\vec{\alpha}\,(r) = \vec{\alpha} + \Gamma^{-1}(r) \iint\limits_{G(r)} \eta(s,t)\,\xi\,(s,t)\,\vec{\varphi}\,(s,t)\,ds\,dt.$$

The assertion of the theorem is proved if

$$\Gamma^{-1}(r) \iint\limits_{G(r)} \eta(s,t)\,\xi\,(s,t)\,\vec{\varphi}\,(s,t)\,ds\,dt \qquad (3.70)$$

converges to 0 with probability 1 as $r \to \infty$. One has

$$\Gamma^{-1}(r) \iint\limits_{G(r)} \eta(s,t)\,\xi(s,t)\,\vec{\varphi}\,(s,t)\,ds\,dt =$$

$$= \; Q^{-1/2}(r)\left[Q^{1/2}(r)\,\Gamma^{-1\prime}(r)\,Q^{1/2}(r)\right]Q^{-1/2}(r) \times$$

$$\times \iint\limits_{G(r)} \eta(s,t)\,\xi(s,t)\,\vec{\varphi}\,(s,t)\,ds\,dt. \tag{3.71}$$

Denote $\gamma_{jk}(r)$ an element of the matrix $Q^{1/2}(r)\,\Gamma^{-1\prime}(r)\,Q^{1/2}(r)$. By virtue of Lemma 24

$$P\left\{\lim_{r\to\infty}\gamma_{jk}(r)=\gamma_{jk}\right\}=1,$$

where $(\gamma_{jk})_{j,k=1}^{m}=\Gamma^{-1}$. The j-th element of (3.70) is equal to

$$\frac{1}{q_j(r)}\sum_{k=1}^{m}\gamma_{jk}(r)\iint\limits_{G(r)}\eta(s,t)\,\xi(s,t)\,\varphi_k^{\tau}(s,t)\,ds\,dt.$$

Then it is sufficient to show that

$$P\left\{\lim_{r\to\infty}\frac{1}{q_j(r)}\iint\limits_{G(r)}\eta(s,t)\,\xi(s,t)\,\varphi_k^{\tau}(s,t)\,ds\,dt=0\right\}=1.$$

We obtain

$$E\left\{\frac{1}{q_j(r)}\iint\limits_{G(r)}\eta(s,t)\,\xi(s,t)\,\varphi_k^{\tau}(s,t)\,ds\,dt\right\}^2 \le$$

$$\le \frac{1}{q_j^2(r)}\iint\limits_{G(r)}\iint\limits_{G(r)}\varphi_k^{\tau}(s,t)\,\varphi_k^{\tau}(\tilde{s},\tilde{t})\,R_1\left(\sqrt{(s-\tilde{s})^2+(t-\tilde{t})^2}\right)\times$$

$$\times \left[R \left(\sqrt{(s-\widetilde{s})^2 + (t-\widetilde{t})^2} \right) + a^2 \right] ds \, dt \, d\widetilde{s} \, d\widetilde{t} \le \frac{L}{q_j^2(r)}.$$

By condition 1a we have $q_j(r) \ge L \, r$. Thus,

$$E \left\{ \frac{1}{q_j(r)} \iint\limits_{G(r)} \eta(s,t) \, \xi(s,t) \, \varphi_k^r(s,t) \, ds \, dt \right\}^2 \le \frac{\widetilde{L}}{r^2}.$$

Analogously to the proof of Lemma 24 it can be shown that

$$E \left\{ \max_{r_1 < r < r_2} \left[\frac{1}{q_j(r)} \iint\limits_{G(r_1,r)} \eta(s,t) \, \xi(s,t) \, \varphi_k^r(s,t) \, ds \, dt \right] \right\}^2 \le \frac{L(r_2^2 - r_1^2)}{q_j(r_1) \, r_1^2}.$$

The rest of the proof is analogous to the proof of Lemma 24. The theorem is proved.

Denote

$$H = \int\limits_0^{2\pi} \int\limits_0^\infty r \, R_1(r) \, \left[R(r) + a^2 \right] B \left(r \cos \varphi, r \sin \varphi \right) dr \, d\varphi.$$

The following assertion is valid.

Theorem 38 *Suppose that conditions* $1-5$ *are satisfied. Then the distribution of the vector* $Q^{1/2}(r) \left[\overrightarrow{\alpha}(r) - \overrightarrow{\alpha} \right]$ *converges weakly as* $r \to \infty$ *to Gaussian distribution in* $\mathcal{R}^m$ *with mean 0 and the variance matrix* $F = \Gamma^{-1} H \Gamma^{-1}$.

Proof. It follows from conditions 2 and 3 that [61], [76]

$$E \left| \eta_1 \eta_2 - E \eta_1 E \eta_2 \right| \le (4 + 6 c) \psi^{\frac{\delta}{2+\delta}} (d(S,F)), \qquad (3.72)$$

where η_1 and η_2 are measurable relatively to σ-algebras $\mathcal{L}(S)$ and $\mathcal{L}(F)$, and $\mathcal{L}(S)$ is the σ-algebra generated by the random field $\eta(s,t)$, $(s,t) \in S$. In (3.72) $d(S,F)$ is the distance between sets S and F, $\psi(d)$ is the strong mixing condition coefficient. It follows from (3.72) that if $\xi(s,t)$ and $\eta(s,t)$ satisfy conditions $1-4$ and $a_r(s,t)$ satisfies the relationship

$$\lim_{r\to\infty} E \left[\iint\limits_{G(r)} a_r(s,t)\, \xi(s,t)\, \eta(s,t)\, ds\, dt \right]^2 = \sigma^2 > 0$$

then the random variable

$$\zeta_r = \iint\limits_{G(r)} a_r(s,t)\, \xi(s,t)\, \eta(s,t)\, ds\, dt$$

is asymptotically normal with parameters $(0,\sigma^2)$ [61], [76].

Let us consider the vector

$$Q^{1/2}(r) \left[\vec{\alpha}\,(r) - \vec{\alpha} \right] = Q^{1/2}(r)\, \Gamma^{-1}(r)\, Q^{1/2}(r) \times$$

$$\times Q^{-1/2}(r) \iint\limits_{G(r)} \xi(s,t)\, \eta(s,t)\, \vec{\varphi}\,(s,t)\, ds\, dt.$$

By Lemma 24

$$P \left\{ \lim_{r\to\infty} Q^{1/2}(r)\, \Gamma^{-1}(r)\, Q^{1/2}(r) = \Gamma^{-1} \right\} = 1.$$

Thus, it is sufficient to consider the limit distribution of the vector

$$\vec{\zeta}\,(r) = \iint\limits_{G(r)} \xi(s,t)\, \eta(s,t)\, \vec{\varphi}^{\,r}\,(s,t)\, ds\, dt.$$

Evidently, $E\,\vec{\zeta}\,(r) = \vec{0}$. Let us find $E\,\vec{\zeta}\,(r)\,\vec{\zeta}'\,(r)$:

$$E\,\vec{\zeta}\,(r)\,\vec{\zeta}'\,(r) =$$

$$= E \iint\limits_{G(r)} \iint\limits_{G(r)} \xi(s,t)\, \xi(\tilde{s},\tilde{t})\, \eta(s,t)\, \eta(\tilde{s},\tilde{t})\, \vec{\varphi}^{\,r}\,(s,t)\, \vec{\varphi}^{\,r'}\,(\tilde{s},\tilde{t})\, ds\, dt\, d\tilde{s}\, d\tilde{t} =$$

$$= \iint\limits_{G(r)} \ \iint\limits_{G(r)} R_1\left(\sqrt{(s-\tilde{s})^2+(t-\tilde{t})^2}\right)\left[R\left(\sqrt{(s-\tilde{s})^2+(t-\tilde{t})^2}\right)+a^2\right] \times$$

$$\times \ \vec{\varphi^r}(s,t)\ \vec{\varphi^r}'(\tilde{s},\tilde{t})\,ds\,dt\,d\tilde{s}\,d\tilde{t} = \iint\limits_{G(r)} \ \iint\limits_{(s-\tilde{s})^2+(t-\tilde{t})^2\le r^2} R_1\left(\sqrt{\tilde{s}^2+\tilde{t}^2}\right) \times$$

$$\times \ \left[R\left(\sqrt{\tilde{s}^2+\tilde{t}^2}\right)+a^2\right]\vec{\varphi^r}(s,t)\ \vec{\varphi^r}'(s+\tilde{s},t+\tilde{t})\,ds\,dt\,d\tilde{s}\,d\tilde{t}.$$

The following relationship is valid for the internal integral:

$$\iint\limits_{(s-\tilde{s})^2+(t-\tilde{t})^2\le r^2} \vec{\varphi^r}(s,t)\ \vec{\varphi^r}'(s+\tilde{s},t+\tilde{t})\,ds\,dt \ \to B(s,t), \quad r\to\infty.$$

Using Lebesgue theorem of limit transition under an integral sign analogously to Lemma 18 one has

$$\iint\limits_{G(r)} \ \iint\limits_{(s-\tilde{s})^2+(t-\tilde{t})^2\le r^2} R_1\left(\sqrt{\tilde{s}^2+\tilde{t}^2}\right)\left[R\left(\sqrt{\tilde{s}^2+\tilde{t}^2}\right)+a^2\right] \times$$

$$\times \ \vec{\varphi^r}(s,t)\ \vec{\varphi^r}'(s+\tilde{s},t+\tilde{t})\,ds\,dt\,d\tilde{s}\,d\tilde{t} \to$$

$$\to \int\limits_0^{2\pi}\int\limits_0^{\infty} r\,R_1(r)\left[R(r)+a^2\right]B\left(r\cos\varphi,\,r\sin\varphi\right)dr\,d\varphi.$$

The theorem is proved.

Remark. *The condition that the random field $\xi(s,t)$ in Gaussian in Lemma 24 and Theorems 37, 38 can be omitted if we demand some additional restrictions on the moments of $\xi(s,t)$ to be fulfilled.*

3.6 GAUSSIAN REGRESSION MODELS FOR QUASISTATIONARY RANDOM PROCESSES

Let $\xi(t)$ be a random process with orthogonal increments, $E\,\xi(t) = 0$ for every t. Nondecreasing function $F(t)$ concerning with $\xi(t)$ via the relationship

$$E\left[\xi(t) - \xi(u)\right]^2 = F(t) - F(u), \quad t \geq u$$

responds to the process $\xi(t)$.

Suppose that S is a set of functions $f(t)$ with the property

$$\int\limits_{-\infty}^{\infty} |f(t)|^2 \, d\xi(t) < \infty.$$

We can define the stochastic integral [31]

$$\int\limits_{-\infty}^{\infty} f(t)\,d\xi(t), \tag{3.73}$$

and consider it in the mean square convergence sense . It is also evident that for $f(t)$, $g(t) \in S$

$$E \int\limits_{-\infty}^{\infty} f(t)\,d\xi(t) \int\limits_{-\infty}^{\infty} g(t)\,d\xi(t) = E \int\limits_{-\infty}^{\infty} f(t)\,g(t)\,dF(t).$$

Analogously to (3.73) we can define for any $s \in (-\infty, \infty)$ the integral

$$\eta(s) = \int\limits_{-\infty}^{\infty} g(t, s)\,d\xi(t) \tag{3.74}$$

for any function $g(t, s)$ such that

$$\int\limits_{-\infty}^{\infty} \int\limits_{-\infty}^{\infty} |g(t, s)|^2 \, dt\,ds < \infty.$$

It can be shown [31] that for each s the stochastic integral (3.74) can be defined as a function of s in such a way that the random process $\eta(s)$ will be measurable.

Let us consider the following model of observations:

$$S_{\theta_0}(t) = m(\theta_0, t) + \int_{-\infty}^{\infty} h(t, \tilde{\tau}) \, d\xi(\tilde{\tau}), \qquad (3.75)$$

where $\xi(t)$ is the random process with orthogonal increments defined above, θ_0 is an unknown parameter belonging to Θ, $m(\theta, t)$ is a known function, $h(t, \tau')$ satisfies the condition

$$\int_{-\infty}^{\infty} \int_{-\infty}^{\infty} |h(t, s)|^2 \, dF(s) \, dt < \infty. \qquad (3.76)$$

Definition 2 [99]. *We said a random process $\alpha(t)$ to be quasistationary if the following conditions are fulfilled:*

1. *$E \, \alpha(t) = a(t)$, $|a(t)| \leq C$;*

2. *$E \, \alpha(t) \, \alpha(\tau) = R(t, \tau)$, $|R(t, \tau)| \leq c$;*

3. *$\lim_{T \to \infty} \dfrac{1}{T} \displaystyle\int_0^T R(t, t - \tau) \, dt = R(\tau)$ for all τ.*

The following assertion is valid.

Theorem 39 *Let for any $\theta \in \Theta$ the process $S_\theta(t)$ defined by (3.75) be quasistationary. Then as $T \to \infty$ with probability 1*

$$\sup_\theta \left| \frac{1}{T} \int_0^T [S_\theta(t) \, S_\theta(s) - E \, S_\theta(t) \, S_\theta(s)] \, dt \right| \to 0, \qquad (3.77)$$

$$\sup_\theta \left| \frac{1}{T} \int_0^T [S_\theta(t) \, m_\theta(s) - E \, S_\theta(t) \, m_\theta(s)] \, dt \right| \to 0, \qquad (3.78)$$

$$\sup_{\theta} \left| \frac{1}{T} \int_0^T [S_\theta(t)\,V(s) - E\,S_\theta(t)\,V(s)]\,dt \right| \to 0, \qquad (3.79)$$

where $V(s) = \int_{-\infty}^{\infty} h(t,s)\,d\xi(t)$ and $\xi(t)$ is Gaussian process with orthogonal increments.

At first we shall prove the following auxiliary statement.

Lemma 25 *Suppose that* $\sup_{\theta} |m(\theta, t)| \le c_m$ *for all* t. *Then*

$$E \left| \int_r^T [V^2(t) - E\,V^2(t)]\,dt \right|^2 \le c\,(T - r), \quad 0 < r < T; \qquad (3.80)$$

$$E \sup_{\theta} \left| \int_r^T [V(t)\,m(\theta, t)]\,dt \right|^2 \le c\,(T - r), \quad 0 < r < T. \qquad (3.81)$$

Proof. For proving (3.80) one has

$$E \left| \int_r^T [V^2(t) - E\,V^2(t)]\,dt \right|^2 = E \left| \int_r^T \left\{ \left[\int_{-\infty}^{\infty} h(s,t)\,d\xi(s) \right]^2 - \right.\right.$$

$$\left.\left. - \int_{-\infty}^{\infty} h^2(s,t)\,dF(s) \right\} dt \right|^2 = E \int_r^T \int_r^T \left[\int_{-\infty}^{\infty} h(s,t)\,d\xi(s) \right]^2 \times$$

$$\times \left[\int_{-\infty}^{\infty} h(\tilde{s},\tilde{t})\,d\xi(\tilde{s}) \right]^2 dt\,d\tilde{t} - 2\,E \int_r^T \left[\int_{-\infty}^{\infty} h(s,t)\,d\xi(s) \right]^2 dt \times$$

$$\times \int_{-\infty}^{\infty} h^2(\tilde{s},\tilde{t})\, dF(\tilde{s})\, d\tilde{t} + \int_{r}^{T}\int_{r}^{T}\int_{-\infty}^{\infty} h^2(s,t)\, dF(s)\, dt \int_{-\infty}^{\infty} h^2(\tilde{s},\tilde{t})\, dF(\tilde{s})\, d\tilde{t} =$$

$$= 2 \int_{r}^{T}\int_{r}^{T}\int_{-\infty}^{\infty} h(s,t)\, h(\tilde{s},\tilde{t})\, dF(s)\, dt\, d\tilde{t} \le$$

$$\le 2 \left[\int_{r}^{T}\int_{r}^{T}\int_{-\infty}^{\infty} h^2(s,t)\, dF(s)\, dt\, d\tilde{t} \int_{r}^{T}\int_{r}^{T}\int_{-\infty}^{\infty} h^2(\tilde{s},\tilde{t})\, dF(s)\, dt\, d\tilde{t} \right]^{1/2} \le$$

$$\le c(T-r)^{1/2}\, (T-r)^{1/2} = c\,(T-r).$$

Now for (3.81)

$$E \sup_{\theta} \left| \int_{r}^{T} V(t)\, m(\theta,t)\, dt \right|^2 =$$

$$= E \sup_{\theta} \left| \int_{r}^{T}\int_{-\infty}^{\infty} h(s,t)\, d\xi(s)\, m(\theta,t)\, dt \right|^2 \le$$

$$\le E \left\{ \int_{r}^{T} \left| \int_{-\infty}^{\infty} h(s,t)\, d\xi(s) \right| \left| \sup_{\theta} m(\theta,t) \right| dt \right\}^2 \le$$

$$\le c_m^2\, E \left\{ \int_{r}^{T} \left| \int_{-\infty}^{\infty} h(s,t)\, d\xi(s) \right|^2 dt \right\} (T-r) =$$

$$= c_m^2 \left\{ \int_{r}^{T}\int_{-\infty}^{\infty} h^2(s,t)\, dF(s)\, dt \right\} (T-r) \le c\,(T-r).$$

The lemma is proved.

Lemma 26 *Let*

$$R_r^T = \sup_\theta \left| \int_r^T \left[S_\theta^2(t) - E S_\theta^2(t) \right] dt \right|.$$

Then $E \, | \, R_r^T \, |^2 \le c \, (T - r)$.

Proof. One has

$$E \sup_\theta \left| \int_r^T \left[S_\theta^2(t) - E S_\theta^2(t) \right] dt \right|^2 =$$

$$= E \sup_\theta \left| \int_r^T \left\{ \left[m(\theta,t) + \int_{-\infty}^\infty h(t,\tau) \, d\xi(\tau) \right]^2 - \right. \right.$$

$$\left. \left. - E \left[m(\theta,t) + \int_{-\infty}^\infty h(t,\tau) \, d\xi(\tau) \right]^2 \right\} dt \right|^2 = E \sup_\theta \left| \int_r^T \left\{ m^2(\theta,t) + 2m(\theta,t) \times \right. \right.$$

$$\times \int_{-\infty}^\infty h(t,\tau) \, d\xi(\tau) + \left(\int_{-\infty}^\infty h(t,\tau) \, d\xi(\tau) \right)^2 - m^2(\theta,t) -$$

$$\left. - \int_{-\infty}^\infty h^2(t,\tau) \, dF(\tau) \right\} dt \right|^2 \le E \int_r^T \int_r^T \left(\int_{-\infty}^\infty h(t,\tau) \, d\xi(\tau) \right)^2 \times$$

$$\times \left(\int_{-\infty}^\infty h(\widetilde{t},\widetilde{\tau}) \, d\xi(\widetilde{\tau}) \right)^2 dt \, d\widetilde{t} - 2 \, E \int_r^T \int_r^T \left(\int_{-\infty}^\infty h(t,\tau) \, d\xi(\tau) \right)^2 dt \times$$

$$\times \int_{-\infty}^\infty h^2(\widetilde{t},\widetilde{\tau}) \, dF(\widetilde{\tau}) \, d\widetilde{t} + \left[\int_r^T \int_{-\infty}^\infty h^2(t,\tau) \, dt \, dF(t) \right]^2 + 2 \, c_m^2 \, (T - r) \times$$

$$\times \int\limits_r^T \int\limits_{-\infty}^{\infty} h^2(t,\tau)\, dF(\tau)\, dt \leq \int\limits_r^T \int\limits_r^T \int\limits_{-\infty}^{\infty} h^2(t,\tau)\, dF(\tau) \int\limits_{-\infty}^{\infty} h^2(\tilde{t},\tilde{\tau})\, dF(\tilde{\tau})\, dt\, d\tilde{t} +$$

$$+ \; 2 \int\limits_r^T \int\limits_r^T \left[\int\limits_{-\infty}^{\infty} h(t,\tau)\, h(\tilde{t},\tau)\, dF(\tau) \right]^2 dt\, d\tilde{t} - 2 \left(\int\limits_r^T \int\limits_{-\infty}^{\infty} h^2(t,\tau)\, dF(\tau)\, dt \right)^2 +$$

$$+ \; 4\, c_m^2\, (T-r) \int\limits_r^T \int\limits_{-\infty}^{\infty} h^2(t,\tau)\, dF(\tau)\, dt \leq \left[\int\limits_r^T \int\limits_{-\infty}^{\infty} h^2(t,\tau)\, dF(\tau)\, dt \right]^2 +$$

$$+ \; 2 \int\limits_r^T \int\limits_r^T \int\limits_{-\infty}^{\infty} h^2(t,\tau)\, dF(\tau) \int\limits_{-\infty}^{\infty} h^2(\tilde{t},\tilde{\tau})\, dF(\tilde{\tau})\, dt\, d\tilde{t} -$$

$$- \; 2 \left(\int\limits_r^T \int\limits_{-\infty}^{\infty} h^2(t,\tau)\, dF(\tau)\, dt \right)^2 + 4\, c_m^2\, (T-r) \int\limits_r^T \int\limits_{-\infty}^{\infty} h^2(t,\tau)\, dF(\tau)\, dt \leq$$

$$\leq \left[\int\limits_r^T \int\limits_{-\infty}^{\infty} h^2(t,\tau)\, dF(\tau)\, dt \right]^2 + 4\, c_m^2\, (T-r) \int\limits_r^T \int\limits_{-\infty}^{\infty} h^2(t,\tau)\, dF(\tau)\, dt \leq \tilde{c}\,(T-r).$$

The lemma is proved.

Let us now prove Theorem 39.

Proof of Theorem 39. Assume that $T = N^2$. Then

$$E \left[\frac{1}{N^2} R_0^{N^2} \right]^2 \leq \frac{1}{N^4} c\, N^2 = \frac{c}{N^2}.$$

By Chebyshev inequality

$$P \left\{ \frac{1}{N^2} R_0^{N^2} > \varepsilon \right\} \leq \frac{1}{\varepsilon^2} E \left(R_0^{N^2} \right)^2.$$

Consequently,

$$\sum_{k=1}^{N} P \left\{ \frac{1}{k^2} R_0^{k^2} > \varepsilon \right\} < c \sum_{k=1}^{N} \frac{1}{k^2} < \infty.$$

Thus, by Borel-Cantelli lemma

$$\frac{1}{k^2} R_0^{k^2} \to 0, \quad k \to \infty$$

with probability 1.

Let for $T = T_N$ and $\theta = \theta_N$ the value $\sup\limits_{N^2 \le T \le (N+1)^2} \frac{1}{T} R_0^T$ be found. Then

$$\sup\limits_{N^2 \le T \le (N+1)^2} \frac{1}{T} R_0^T = \frac{1}{T_N} \left| \int_0^{T_N} [S_{\theta_N}^2(t) - E\, S_{\theta_N}^2(t)]\, dt \right| \le$$

$$\le \frac{1}{T_N} \left| \int_0^{N^2} [S_{\theta_N}^2(t) - E\, S_{\theta_N}^2(t)]\, dt \right| +$$

$$+ \frac{1}{T_N} \left| \int_{N^2}^{T_N} [S_{\theta_N}^2(t) - E\, S_{\theta_N}^2(t)]\, dt \right| \le$$

$$\le \frac{1}{N^2} R_0^{N^2} + \frac{1}{N^2} R_{N^2}^{T_N}. \tag{3.82}$$

By Borel-Cantelli lemma the first term of (3.82) tends to 0 with probability 1. For the second term because of Lemma 26 one has

$$E \left| \frac{1}{N^2} R_{N^2}^{T_N} \right|^2 \le \frac{c}{N^4} [(N+1)^2 - N^2] \le \frac{c}{N^3}.$$

Consequently, using Borel-Cantelli lemma we obtain

$$\sup\limits_{N^2 \le T \le (N+1)^2} \frac{1}{T} R_0^T \to 0 \quad \text{as} \quad N \to \infty$$

with probability 1.

Relations (3.78) and (3.79) are proved analogously. The proof is complete.

Corollary 2 *It follows from Theorem 39 that*

$$\sup\limits_{\theta} \left| \frac{1}{T} \int_0^T [V(t)\, m(\theta, t) - E\, V(t)\, m(\theta, t)]\, dt \right| \to 0$$

as $T \to \infty$ with probability 1.

We shall note that we demanded the process $\xi(t)$ to be Gaussian for symplifying calculations. This restriction can be removed under additional conditions on moments of $\xi(t)$.

Let us consider now model (3.75). The problem is to estimate the unknown parameter $\theta_0 \in \Theta$. As criterium we will consider the least squares method, where

$$Q_T(\theta) = \frac{1}{T} \int_0^T [S_{\theta_0}(t) - m(\theta, t)]^2 \, dt,$$

and the value θ_T satisfying the relationship

$$\min_\theta Q_T(\theta) = Q_T(\theta_T)$$

is called an estimate of θ_0. The following assertion is valid.

Theorem 40 *Suppose that there exists*

$$\lim_{T \to \infty} \frac{1}{T} \int_0^T [m(\theta, t) - m(\theta_0, t)]^2 \, dt = V(\theta)$$

and conditions of Theorem 39 are satisfied. Then

$$\lim_{T \to \infty} \sup_\theta \left| \frac{1}{T} \int_0^T [S_{\theta_0}(t) - m(\theta, t)]^2 \, dt - V(\theta) \right| = 0.$$

Proof. It is evident that

$$\sup_\theta \left| \frac{1}{T} \int_0^T \left\{ [S_{\theta_0}(t) - m(\theta, t)]^2 \, dt - E\,[S_{\theta_0}(t) - m(\theta, t)]^2 \right\} dt \right| \le$$

$$\le \left| \frac{1}{T} \int_0^T [S_{\theta_0}^2(t) - E\,S_{\theta_0}^2(t)] \, dt \right| + \sup_\theta \left| \frac{2}{T} \int_0^T [S_{\theta_0}(t)\,m(\theta, t) - \right.$$

$$- \ E \, S_{\theta_0}(t) \, m(\theta, t)] \ dt \Big| \to 0, \quad T \to \infty.$$

(3.83)

From (3.83) it follows that the assertion of the theorem is true.

Theorem 41 *Assume that conditions of Theorem 40 and the following conditions are fulfilled:*

1) $m(\theta, t) \neq m(\theta_0, t)$ *for* $\theta \neq \theta_0$;

2) $|m(\theta_1, t) - m(\theta_2, t)| \leq c \, |\theta_1 - \theta_2|$, *where the constant* c *does not depend on* t.

Then

$$P \left\{ \lim_{T \to \infty} \theta_T = \theta_0 \right\} = 1.$$

(3.84)

Proof. We will use Theorem 7. Condition 3) of Theorem 7 is satisfied with the function $\Phi(\theta; \theta_0) = V(\theta)$ because of Theorem 40. Let us check condition 4) of Theorem 7. Taking into account (3.76) one has

$$\zeta_T(\gamma, \overline{\theta}) = \sup_{|\theta - \overline{\theta}| < \gamma} |Q_T(\theta) - Q_T(\overline{\theta})| \leq \sup_{|\theta - \overline{\theta}| < \gamma} \left| \frac{1}{T} \int_0^T \left[m(\theta, t) - m(\overline{\theta}, t) \right]^2 dt \right| +$$

$$+ \sup_{|\theta - \overline{\theta}| < \gamma} \left| \frac{1}{T} \int_0^T \left[m(\theta, t) - m(\overline{\theta}, t) \right] \int_{-\infty}^{\infty} h(t, \tau) \, d\xi(\tau) \, dt \right| \leq$$

$$\leq c \gamma \left[1 + \frac{1}{T} \int_0^T \left| \int_{-\infty}^{\infty} h(t, \tau) \, d\xi(\tau) \right| dt \right] \leq \widetilde{c} \gamma$$

with probability 1.

Thus, the conditions of Theorem 7 are fulfilled and this fact implies (3.84). The theorem is proved.

PERIODOGRAM ESTIMATES FOR RANDOM PROCESSES AND FIELDS

In this chapter the periodogram estimates of two types are considered. It is proved that the empirical estimates of the parameter converge to its true value with probability 1, and the speed of the convergence is appreciated. Theorems about the strong consistency of the coefficients estimates in the models are proved. The conditions of the asymptotic normality of the estimates are found.

At the end of the chapter the periodogram estimates for the multi-dimensional parameter are investigated.

4.1   PRELIMINARY RESULTS

We suppose that all random functions, being taken into consideration, are homogeneous in a strict sense, real, mean square continuous, measurable and separable. These restrictions are rather general because any mean square continuous random function with the domain and the set of values which are finite-dimension Euclidian space, is stochastically equivalent to a separable and measurable function [31].

Lemma 27 *[54], [116]. Suppose that the following conditions are fulfilled:*

1) $\{n(t)\} = \Big\{n(t), t \in \mathcal{R}\Big\}$ *is stationary in a strict sense random process satisfying the strong mixing condition with the coefficient*

$$\psi(\tau) \le \frac{c}{1 + \tau^{1+\varepsilon}}, \quad c > 0, \quad \varepsilon > 0,$$

and $En(t) = 0$;

2) *for some* $\delta > \dfrac{4}{\varepsilon}$

$$E\,|n(t)|^{2+\delta} < \infty.$$

Then for the correlation function $r(t)$ *of the random process* $n(t)$ *we have*

$$|r(t)| \leq \frac{c}{1+t^{1+\varepsilon_1}}, \quad c > 0, \quad \varepsilon_1 > 0, \quad t \geq 0.$$

Lemma 28 [96], [97]. *Let the following conditions be satisfied:*

1) $\left\{ n(\vec{t}) \right\} = \left\{ n(\vec{t}),\ \vec{t} \in \mathcal{R}^m \right\}$ *is homogeneous in a strict sense random field satisfying strong mixing condition;*

2) *the real random variables* ξ *and* η *are measurable relatively to* σ-*algebras* $\mathcal{A}(\mathcal{S})$ *and* $\mathcal{A}(\mathcal{F})$ *respectively, where* $\mathcal{A}(\mathcal{S}) = \sigma\left\{ n(\vec{t}),\ \vec{t} \in \mathcal{S} \right\},\ \mathcal{S} \subset \mathcal{R}^m$;

3) $E\,|\xi|^{2+\delta} < c,\ E\,|\eta|^{2+\delta} < c$ *for some* $\delta > 0, 0 < c < \infty$.

Then

$$E\,|\xi\eta - E\xi\eta| \leq (4+6c)\,\psi^{\frac{\delta}{2+\delta}}\,(d(\mathcal{S}, \mathcal{F})),$$

where ψ *is a function from the strong mixing condition,*

$$d(\mathcal{S}, \mathcal{F}) = \inf\left\{ \|\,\vec{x} - \vec{y}\,\|,\ \vec{x} \in \mathcal{S},\ \vec{y} \in \mathcal{F} \right\}.$$

Lemma 29 [96], [97]. *Assume that the following conditions are fulfilled:*

1) $\left\{ n(\vec{t}) \right\} = \left\{ n(\vec{t}),\ \vec{t} \in \mathcal{R}^m \right\}$ *is homogeneous in a strict sense random field satisfying strong mixing condition with*

$$\psi(d) \leq \frac{c}{1+d^{m+\varepsilon}}, \quad \varepsilon > 0, \quad 0 < c < \infty,$$

and $En(\vec{t}) = 0$;

2) for some $\delta > \dfrac{4m}{\varepsilon}$

$$E\left|n(\overrightarrow{t})\right|^{2+\delta} < \infty.$$

Then for correlation function $r(\overrightarrow{t})$ of the random field $n(\overrightarrow{t})$ we have

$$\left|r(\overrightarrow{t})\right| \leq \dfrac{c}{1 + \|\overrightarrow{t}\|^{m+\varepsilon_1}}, \quad 0 < c < \infty, \quad \varepsilon_1 > 0.$$

We will use integrals of the type

$$I_A(\overrightarrow{\lambda}) = \int_A n(\overrightarrow{t})\,\varphi(\overrightarrow{t}, \overrightarrow{\lambda})\,d\overrightarrow{t}, \tag{4.1}$$

where A is some measurable in Lebesgue sense bounded set from $\mathcal{R}^m$, $m \geq 1$, $n(\overrightarrow{t})$ is the measurable separable random function with

$$\int_A E\left|n(\overrightarrow{t})\right| d\overrightarrow{t} < \infty,$$

$\varphi(\overrightarrow{t}, \overrightarrow{\lambda})$ is the deterministic continuous function. Because of Fubini theorem the integral $I_A(\overrightarrow{\lambda})$ exists with probability 1 for any $\overrightarrow{\lambda} \in \mathcal{R}^m$ and it is a continuous function almost surely [118].

Lemma 30 *Let the following conditions be fulfilled:*

1) $\left\{n(\overrightarrow{t})\right\} = \left\{n(\overrightarrow{t}),\ \overrightarrow{t} \in \mathcal{R}^m\right\}$ *is homogeneous in a strict sense random field satisfying the strong mixing condition with*

$$\psi(d) \leq \dfrac{c}{1 + d^{m+\varepsilon}}, \quad 0 < c < \infty, \quad \varepsilon > 0, \quad d \geq 0,$$

and $En(\overrightarrow{t}) = 0$;

2) *for some $\delta > \dfrac{4m}{\varepsilon}$*

$$E\left|n(\overrightarrow{t})\right|^{4+\delta} < \infty.$$

Then

$$\sup_{\vec{\omega} \in \mathcal{R}^m} \left| \frac{1}{T^m} \int_{[0,T]^m} n(\vec{t}) e^{i(\vec{\omega}, \vec{t})} d\vec{t} \right| \to 0, \quad T \to \infty \qquad (4.2)$$

with probability 1, where

$$(\vec{\omega}, \vec{t}) = \sum_{i=1}^{m} \omega_i t_i, \quad \vec{\omega} = (\omega_i)_{i=1}^{m}, \quad \vec{t} = (t_i)_{i=1}^{m}.$$

Proof. Denote

$$I_T(\vec{\omega}) = \frac{1}{T^m} \int_{[0,T]^m} n(\vec{t}) e^{i(\vec{\omega}, \vec{t})} d\vec{t}.$$

We have that $n(\vec{t})$ is separable measurable random function satisfying condition 2, and $\varphi(\vec{t}, \vec{\omega}) = e^{i(\vec{\omega}, \vec{t})}$ is continuous function. Then $I_T(\vec{\omega})$ exists with probability 1 for any $\vec{\omega} \in \mathcal{R}^m$, it is almost surely continuous in $\vec{\omega}$ function, and, consequently, $\sup_{\vec{\omega} \in \mathcal{R}^m} \left| I_T(\vec{\omega}) \right|$ is a random variable. The value $\breve{\omega}_T$, for which $\sup_{\vec{\omega} \in \mathcal{R}^m} \left| I_T(\vec{\omega}) \right|$ is reached, is defined with probability 1 because $I_T(\vec{\omega}) \to 0$ almost surely as $\| \vec{\omega} \| \to \infty$. It follows from Theorem 6 that $\breve{\omega}_T$ can be chosen to be random variable.

Let us consider the value

$$\left| I_T(\vec{\omega}) \right|^2 = \frac{1}{T^{2m}} \int_{[0,T]^{2m}} e^{i(\vec{\omega}, \vec{t} - \vec{s})} n(\vec{t}) n(\vec{s}) d\vec{t} \, d\vec{s}.$$

Denote $\vec{t} - \vec{s} = \vec{u}$, $\vec{t} = \vec{v}$, $u^+ = \max(0, u)$, $u^- = \max(0, -u)$. Then

$$\left| I_T(\vec{\omega}) \right|^2 = \frac{1}{T^{2m}} \int_{[-T,T]^m} e^{i(\vec{\omega}, \vec{u})} d\vec{u} \int_{\substack{m \\ \underset{i=1}{\times} [u_i^+, T - u_i^-]}} n(\vec{v}) n(\vec{v} + \vec{u}) d\vec{v} =$$

$$= \frac{1}{T^{2m}} \int_{[-T,T]^m} e^{-i(\vec{\omega}, \vec{u})} d\vec{u} \int_{\substack{m \\ \underset{i=1}{\times} [u_i^-, T - u_i^+]}} n(\vec{v}) n(\vec{v} + \vec{u}) d\vec{v} \leq$$

$$\leq \frac{1}{T^{2m}} \int\limits_{[-T,T]^m} \left| \int\limits_{\substack{m \\ \underset{i=1}{X} [u_i^-, T - u_i^+]}} n(\vec{v})\, n(\vec{v} + \vec{u})\, d\,\vec{v} \right| d\,\vec{u} = I_T,$$

where "X" means Decart product.

Let

$$r_T(\vec{u}) = \frac{1}{T^m} \int\limits_{[0,T]^m} n(\vec{v})\, n(\vec{v} + \vec{u})\, d\,\vec{v},$$

$$\Delta_T = \int\limits_{[0,T]^m} \left| r_T(\vec{u}) \right| d\,\vec{u}.$$

Estimate $E \sup\limits_{\vec{\omega} \in \mathcal{R}^m} \left| I_T(\vec{\omega}) \right|^2$. Consider the value $E\Delta_T$. Evidently

$$E\,\Delta_T = \int\limits_{[0,T]^m} E \left| r_T(\vec{u}) \right| d\,\vec{u} \leq \int\limits_{[0,T]^m} \sqrt{E\, r_T^2(\vec{u})}\, d\,\vec{u}.$$

Let us estimate $E\, r_T^2(\vec{u})$. We have

$$E\, r_T^2(\vec{u}) = \frac{1}{T^{2m}} \int\limits_{[0,T]^{2m}} E \left[n(\vec{u} + \vec{t})\, n(\vec{t})\, d\,\vec{t} \right]^2 =$$

$$= \frac{1}{T^{2m}} \int\limits_{[0,T]^{2m}} E \left[n(\vec{u} + \vec{t})\, n(\vec{t})\, n(\vec{u} + \vec{s})\, n(\vec{s}) \right] d\,\vec{t}\, d\,\vec{s} =$$

$$= \frac{1}{T^{2m}} \int\limits_{[0,T]^{2m}} E \left[n(\vec{0})\, n(\vec{u})\, n(\vec{s} - \vec{t})\, n(\vec{s} - \vec{t} + \vec{u}) \right] d\,\vec{t}\, d\,\vec{s} =$$

$$= \frac{1}{T^m} \int\limits_{[-T,T]^m} \prod_{i=1}^{m} \left(1 - \frac{|t_i|}{T} \right) E \left[n(\vec{0})\, n(\vec{u})\, n(\vec{t})\, n(\vec{u} + \vec{t}) \right] d\,\vec{t} \leq$$

$$\leq \frac{1}{T^{2m}} \int\limits_{[-T,T]^m} \left| E\,n(\vec{0})\,n(\vec{u})\,n(\vec{t})\,n(\vec{u}+\vec{t}) \right| d\,\vec{t}\,.$$

It follows from conditions 1, 2 and Lemmas 28, 29, that

$$\left| E\,n(\vec{0})\,n(\vec{u})\,n(\vec{t})\,n(\vec{u}+\vec{t}) \right| \leq \left| E\,n(\vec{0})\,n(\vec{u})\,n(\vec{t})\,n(\vec{u}+\vec{t}) - r^2(\vec{u}) - \right.$$

$$\left. -r^2(\vec{t}) - r(\vec{t}-\vec{u})\,r(\vec{t}+\vec{u}) \right| + r^2(\vec{u}) + r^2(\vec{t}) + \left| r(\vec{t}-\vec{u})\,r(\vec{t}+\vec{u}) \right| \leq$$

$$\leq \frac{c}{1 + \min\left\{ \|\vec{u}\|, \|\vec{t}\|, \|\vec{u}-\vec{t}\|, \|\vec{u}+\vec{t}\| \right\}^{m+\varepsilon_1}} + c_1 \left[r^2(\vec{u}) + r^2(\vec{t}) + \right.$$

$$\left. + \left| r(\vec{t}-\vec{u})\,r(\vec{t}+\vec{u}) \right| \right], \quad \varepsilon_1 > 0, \quad 0 < c < \infty, \quad 0 < c_1 < \infty.$$

Consequently

$$\frac{1}{T^m} \int\limits_{[0,T]^m} \sqrt{E\,r_T^2(\vec{u})}\; d\,\vec{u} \leq \frac{1}{T^{\frac{m}{2}}} \left(\int\limits_{[0,T]^m} E\,r_T^2(\vec{u})\,d\,\vec{u} \right)^{1/2} \leq \frac{c}{T^{\frac{m}{2}}}$$

with some constant $0 < c < \infty$, which does not depend on T. Analogously we can obtain the same estimate for $E\,I_T$. Then

$$E\sup_{\vec{\omega} \in \mathcal{R}^m} \left| I_T(\vec{\omega}) \right|^2 \leq \frac{c}{T^{\frac{m}{2}}}, \quad 0 < c < \infty.$$

Let us take $T_n = n^3$, $n \geq 1$. Then by virtue of Borel-Cantelli Lemma

$$\sup_{\vec{\omega} \in \mathcal{R}^m} \left| I_{T_n}(\vec{\omega}) \right| \to 0, \quad n \to \infty \tag{4.3}$$

with probability 1. Let

$$\zeta_n = \sup_{T_n \leq T \leq T_{n+1}} \sup_{\vec{\omega} \in \mathcal{R}^m} \left| I_T(\vec{\omega}) - I_{T_n}(\vec{\omega}) \right|.$$

We have

$$\sup_{\vec{\omega} \in \mathcal{R}^m} \left| I_T(\vec{\omega}) \right| \leq \sup_{\vec{\omega} \in \mathcal{R}^m} \left| I_{T_n}(\vec{\omega}) \right| + \zeta_n, \quad T_n \leq T \leq T_{n+1}.$$

Then it is sufficient to prove that $\zeta_n \to 0$, $n \to \infty$ with probability 1. The following inequality takes place for ζ_n:

$$\zeta_n \leq \frac{T_{n+1}^m - T_n^m}{T_n^m} \sup_{\vec{w} \in \mathcal{R}^m} \left| I_{T_n}(\vec{w}) \right| + \frac{1}{T_n^m} \int_{D_{mn}} \left| n(\vec{t}) \right| d\vec{t},$$

where $D_{mn} = [0, T_{n+1}]^m \setminus [0, T_n]^m$. Because of (4.3) the first term converges to 0 as $n \to \infty$ with probability 1. For the second term we have

$$\frac{1}{T^{2m}} E \left[\int_{D_{mn}} \left| n(\vec{t}) \right| d\vec{t} \right]^2 \leq \frac{c(T_{n+1} - T_n) T_{n+1}^{m-1}}{T_n^{2m}} \int_{D_{mn}} E \left| n(\vec{t}) \right|^2 d\vec{t} \leq$$

$$\leq c_1 \frac{(T_{n+1} - T_n)^2 T_{n+1}^{2(m-1)}}{T_n^{2m}} \leq c_2 \left(\frac{T_{n+1} - T_n}{T_n} \right)^2 \leq \frac{c_3}{n^2}, \quad 0 < c, c_1, c_3 < \infty.$$

From this relationship we obtain

$$\frac{1}{T_n^m} \int_{D_{mn}} \left| n(\vec{t}) \right| d\vec{t} \to 0, \quad n \to \infty$$

with probability 1. The lemma is proved.

Lemma 31 *Let conditions of Lemma 30 be fulfilled. Then for any function of the type*

$$\varphi(\vec{t}) = \sum_{k_1,\ldots,k_m = -\infty}^{\infty} c_{k_1,\ldots,k_m} e^{i(\vec{\lambda}_{k_1,\ldots,k_m}, \vec{t})},$$

where $\vec{\lambda}_{k_1,\ldots,k_m}$ is a real vector from $\mathcal{R}^m$ and coefficients $c_{k_1,\ldots,k_m}$ satisfy the condition

$$\sum_{k_1,\ldots,k_m = -\infty}^{\infty} \left| c_{k_1,\ldots,k_m} \right| < \infty, \tag{4.4}$$

the following relationship takes place:

$$P \left\{ \lim_{T \to \infty} \sup_{\vec{w} \in \mathcal{R}^m} \left| \frac{1}{T^m} \int_{[0,T]^m} n(\vec{t}) \varphi(\vec{w} \cdot \vec{t}) d\vec{t} \right| = 0 \right\} = 1,$$

where $(\vec{w} \cdot \vec{t}) = (w_1 t_1, \ldots, w_m t_m)$.

Proof. Denote

$$\tilde{I}_T(\vec{\omega}) = \frac{1}{T^m} \int\limits_{[0,T]^m} n(\vec{t})\,\varphi(\vec{\omega}\cdot\vec{t})\,d\,\vec{t}\,.$$

Let us estimate $\sup\limits_{\vec{\omega}\in\mathcal{R}^m}\left|\tilde{I}_T(\vec{\omega})\right|$. We have

$$\sup\limits_{\vec{\omega}\in\mathcal{R}^m}\left|\tilde{I}_T(\vec{\omega})\right| \le$$

$$\le \sum_{k_1,\ldots,k_m=-\infty}^{\infty} c_{k_1,\ldots,k_m} \sup\limits_{\vec{\omega}\in\mathcal{R}^m} \frac{1}{T^m}\left|\int\limits_{[0,T]^m} n(\vec{t})\,e^{i(\vec{\lambda}_{k_1,\ldots,k_m},\vec{\omega}\cdot\vec{t})}d\,\vec{t}\right| =$$

$$= \sum_{k_1,\ldots,k_m=-\infty}^{\infty} |c_{k_1,\ldots,k_m}| \sup\limits_{\vec{\omega}\in\mathcal{R}^m} \frac{1}{T^m}\left|\int\limits_{[0,T]^m} n(\vec{t})\,e^{i(\vec{\omega},\vec{t})}d\,\vec{t}\right|.$$

Because of condition (4.4) and Lemma 30 we obtain

$$\sup\limits_{\vec{\omega}\in\mathcal{R}^m}\left|\tilde{I}_T(\vec{\omega})\right| \to 0, \quad T\to\infty$$

with probability 1. The lemma is proved.

Let us note that assertion of Lemma 30 for random process with some different conditions on moments of random process $n(t)$ and strong mixing coefficient is contained in [35], [57]. Lemmas 30 and 31 will play the essential role for investigation of asymptotic behavior of estimates being consider below.

For proving the asymptotic normality of estimates we will need a statement about asymptotic normality of integrals

$$\int\limits_0^T a_T(t)\,n(t)\,dt$$

with some function $a_T(t)$, $T > 0$.

Theorem 42 *[35], [44]. Let function $a_T(t)$ satisfy the following conditions:*

1) *$a_T(t)$ is a real measurable function defined for $t > 0$ and such that for each $T > 0$*

$$W^2(T) = \int_0^T a_T^2(t)\, dt < \infty;$$

2) *for some constant $0 < c < \infty$*

$$W^{-1}(T) \sup_{0 \leq t \leq T} |a_T(t)| \leq \frac{c}{\sqrt{T}};$$

3) *for any real u there exists the limit*

$$\lim_{T \to \infty} \frac{1}{W^2(T)} \int_0^T a_T(t + |u|)\, a_T(t)\, dt = \rho(u),$$

and the function $\rho(u)$ is continuous.

Suppose that random process $\{n(t)\}$ satisfies conditions of Lemma 27. Then the value

$$\frac{1}{W(T)} \int_0^T a_T(t)\, n(t)\, dt$$

is asymptotically normal as $T \to \infty$ with parameters 0 and

$$\sigma^2 = 2\pi \int_{-\infty}^{\infty} f(\lambda)\, d\mu(\lambda),$$

where $\mu(\lambda)$ is the monotone nondecreasing function bounded on $\mathcal{R}$, defined from introducing

$$\rho(u) = \int_{-\infty}^{\infty} e^{i\lambda u}\, d\mu(\lambda), \tag{4.5}$$

and $f(\lambda)$ is spectral density of $n(t)$ connected with correlation function $r(t)$ by the relationship

$$r(t) = \int_{-\infty}^{\infty} e^{it\lambda} f(\lambda)\, d\lambda.$$

Let us note that under conitions 1) − 4) representation (4.5) takes place because of Bochner-Hinchin theorem, and continuous bounded on $\mathcal{R}$ spectral density $f(\lambda)$ exists by virtue of Lemma 27.

The next assertion follows from Theorem 42.

Lemma 32 *Let conditions of Lemma 27 be fulfilled and $f(\omega_0) > 0$. Then all random variables*

$$\frac{1}{\sqrt{T}} \int_0^T \cos(\omega_0 t)\, n(t)\, dt, \qquad \frac{1}{\sqrt{T}} \int_0^T \sin(\omega_0 t)\, n(t)\, dt,$$

$$\frac{1}{T^{3/2}} \int_0^T t \cos(\omega_0 t)\, n(t)\, dt, \qquad \frac{1}{T^{3/2}} \int_0^T t \sin(\omega_0 t)\, n(t)\, dt$$

is asymptotically normal as $T \to \infty$ with parameters $(0, \pi f(\omega_0))$ for the first and the second random variables, and $\left(0, \dfrac{\pi}{3} f(\omega_0)\right)$ for the third and the fourth ones.

Let us write models being considered below and make some assumptions about them.

Let $\left\{ n(\vec{t}\,) \right\} = \left\{ n(\vec{t}\,),\ \vec{t} \in \mathcal{R}^m \right\}$ be real homogeneous in a strict sense random field, $E n(\vec{t}\,) = 0$, $\varphi(\vec{t}\,)$ is the real function defined on $\mathcal{R}^m$. We consider the problem of estimating of the parameter $\vec{\omega} = (\omega_{01}, \ldots, \omega_{0m})$ by observations of random field

$$\left\{ x(\vec{t}\,) \right\} = \left\{ x(\vec{t}\,) = A_0 \varphi(\vec{\omega} \cdot \vec{t}\,) + n(\vec{t}\,),\ \vec{t} \in \mathcal{R}^m \right\}, \tag{4.6}$$

where $A_0 > 0$, $\omega_{0i} > 0$, $i = \overline{1, m}$.

We will make some assumptions relatively to random field $\left\{ n(\vec{t}\,),\ \vec{t} \in \mathcal{R}^m \right\}$ and deterministic function $\varphi(\vec{t}\,),\ \vec{t} \in \mathcal{R}^m$.

A1. Suppose that $\left\{ n(\vec{t}\,),\ \vec{t} \in \mathcal{R}^m \right\}$ is real homogeneous in a strict sense random field with $E n(\vec{t}\,) = 0$, $E n(\vec{s} + \vec{t}\,) n(\vec{s}\,) = r(\vec{t}\,)$, satisfying strong mixing condition with $\psi(d) \le \dfrac{c}{1 + d^{m+\varepsilon}}$, $0 < c < \infty$, $d \ge 0$, $\varepsilon > 0$.

Besides, as it was mentioned above the random field $\left\{ n(\vec{t}),\ \vec{t} \in \mathcal{R}^m \right\}$ is continuous in mean squares sense, measurable and separable.

A2. For some $\delta > \dfrac{4m}{\varepsilon}$

$$ E \left| n(\vec{t}) \right|^{4+\delta} < \infty. $$

A3. Suppose that $\varphi(\vec{t})$ is the real function with the view

$$ \varphi(\vec{t}) = \sum_{k_1,\ldots,k_m=-\infty}^{\infty} c_{k_1,\ldots,k_m}\, e^{i(\vec{\lambda}_{k_1,\ldots,k_m},\ \vec{t})}, $$

$$ \vec{\lambda}_{k_1,\ldots,k_m} = (\lambda_{k_1},\ldots,\lambda_{k_m}) \in \mathcal{R}^m, \qquad \vec{t} = (t_1,\ldots,t_m) \in \mathcal{R}^m, $$

where values $c_{k1,\ldots,k_m}$ and λ_{k_i} satisfy conditions:

$$ \sum_{k_1,\ldots,k_m=-\infty}^{\infty} \left| c_{k_1,\ldots,k_m} \right| < \infty, \qquad \lambda_{k_i} > \lambda_{l_i} > 0 \text{ for } k_i > l_i > 0, \qquad i = \overline{1,m}, $$

$$ \lambda_{k_i} = -\lambda_{-k_i}, \qquad \left| \lambda_{k_i} - \lambda_{l_i} \right| \ge \Delta > 0, \qquad k_i \ne l_i, \qquad i = \overline{1,m}. $$

Because we will consider the case $m = 1$, let us rewrite conditions $A1 - A3$ for this case.

A4. Suppose that $\{n(t)\} = \left\{ n(t),\ t \in \mathcal{R} \right\}$ is real stationary in a strict sense continuous in mean squares sense, measurable random process, $E\, n(t) = 0$, $E\, n(s+t)\, n(s) = r(t)$, satisfying strong mixing condition with $\psi(\tau) \le \dfrac{c}{1 + \tau^{m+\varepsilon}}$, $0 < c < \infty$, $\tau > 0$, $\varepsilon > 0$.

A5. For some $\delta > \dfrac{4}{\varepsilon}$

$$ E\, |n(t)|^{4+\delta} < \infty. $$

A6. Assume that $\varphi(t)$ is the real function with the representation

$$\varphi(t) = \sum_{k=-\infty}^{\infty} c_k \, e^{i\lambda_k t}, \quad \lambda_k \in \mathcal{R},$$

where coefficients c_k and values λ_k satisfy conditions

$$\sum_{k=-\infty}^{\infty} |c_k| < \infty, \quad \lambda_l < \lambda_k \quad \text{for} \quad l > k > 0, \quad c_k = \bar{c}_{-k},$$

$$\lambda_k = -\lambda_{-k}, \quad |\lambda_l - \lambda_k| \geq \Delta > 0 \quad \text{for} \quad l \neq k.$$

We will consider estimates of two types depending on the method of giving a functional for finding the estimates. These types are similar in their properties, but there exist some differences between them. The choice of estimate depends on the specific form of the function $\varphi(t)$.

4.2   ASYMPTOTIC BEHAVIOR OF PERIODOGRAM ESTIMATES OF THE FIRST TYPE

We consider the functional

$$Q_T(\omega) = \left| \frac{1}{T} \int_0^T x(t) \, e^{i\omega t} \, dt \right|^2. \tag{4.7}$$

Suppose that $\omega_T \geq 0$ is such that $Q_T(\omega_T)$ is maximal. We shall note that under assumptions $A4 - A6$ the quantity ω_T is defined with probability 1 and can be chosen measurable because with probability 1 $Q_T(\omega)$ is continuous function of ω, $Q_T(\omega) \to 0$, $\omega \to \infty$, and as for Lemma 30 all conditions of Theorem 6 are fulfilled. Numerical methods of finding of global optimum for functionals similar to (4.7) are considered in [113], [125].

Theorem 43 *Let conditions $A4 - A6$ be fulfilled and*

$$|c_{i_0}| > |c_i|, \quad i \neq i_0, \quad i_0 > 0. \tag{4.8}$$

Then

$$\overline{\omega}_T = \frac{\omega_T}{\lambda_{i_0}} \to \omega_0, \quad T \to \infty$$

with probability 1.

Proof. Fix ω and consider the behavior of $Q_T(\omega)$ as $T \to \infty$:

$$Q_T(\omega) = \left| \frac{1}{T} \int_0^T x(t)\, e^{i\omega t}\, dt \right|^2 = \left| \frac{1}{T} \int_0^T [A_0\, \varphi(\omega_0 t) + n(t)]\, e^{i\omega t}\, dt \right|^2 =$$

$$= \left| \frac{1}{T} \int_0^T A_0\, \varphi(\omega_0 t)\, e^{i\omega t}\, dt \right|^2 + I_T(\omega),$$

where

$$I_T(\omega) = \frac{1}{T^2} \left| \int_0^T n(t)\, e^{i\omega t}\, dt \right|^2 + 2\, Re \frac{A_0}{T^2} \int_0^T \varphi(\omega_0 t)\, e^{i\omega t}\, dt \int_0^T n(t)\, e^{-i\omega t}\, dt.$$

Because of condition *A6*

$$\frac{1}{T} \left| \int_0^T \varphi(\omega_0 t)\, e^{i\omega t}\, dt \right| \le c, \quad 0 < c < \infty.$$

It follows from Lemma 30 that $\sup_\omega \left| I_T(\omega) \right| \to 0$, $T \to \infty$ with probability 1. Let

$$\Phi_T(\omega_0, \omega) = \frac{A_0}{T} \int_0^T \varphi(\omega_0 t)\, e^{i\omega t}\, dt.$$

The following equation takes place:

$$\Phi_T(\omega_0, \omega) = \sum_{k=-\infty}^{\infty} \frac{A_0}{T} c_k \int_0^T e^{i(\omega_0 \lambda_k + \omega) t}\, dt.$$

Let $0 < \widetilde{\delta} < \dfrac{\Delta\omega_0}{2}$. Suppose that for some k $|\lambda_k \omega_0 + \omega| < \widetilde{\delta}$. Then for any $l \neq k$ we have

$$|\lambda_l \omega_0 + \omega| = \left|(\lambda_k \omega_0 + \omega) + (\lambda_l - \lambda_k)\omega_0\right| >$$

$$> \omega_0 \left|\lambda_l - \lambda_k\right| - \frac{1}{2}\Delta \Big| > \frac{1}{2}\Delta\omega_0 > \widetilde{\delta}. \qquad (4.9)$$

That is why taking in consideration (4.8) for any $\delta > 0$ and $0 < \widetilde{\delta} < \dfrac{\Delta\omega_0}{2}$ with probability 1 we obtain

$$\varlimsup_{n\to\infty} \sup_{\omega\geq 0,\, |\lambda_{i_0}\omega_0-\omega|\geq\delta} Q_T(\omega) \leq \varlimsup_{n\to\infty} \sup_{\omega\geq 0,\, |\lambda_{i_0}\omega_0-\omega|\geq\min(\delta,\widetilde{\delta})} Q_T(\omega) \leq$$

$$\leq \varlimsup_{n\to\infty} \sup_{\omega\geq 0,\, |\lambda_{i_0}\omega_0-\omega|\geq\min(\delta,\widetilde{\delta})} |\Phi_T(\omega_0,\omega)|^2 =$$

$$= \varlimsup_{n\to\infty} \sup_{\omega\geq 0,\, |\lambda_{i_0}\omega_0-\omega|\geq\min(\delta,\widetilde{\delta})} \left|\frac{A_0}{T} \sum_{k=-\infty}^{\infty} |c_k| \int_0^T e^{i(\lambda_k,\omega_0+\omega)\,t}\,dt\right|^2 \leq$$

$$\leq A_0^2 \max_{k\neq\pm i_0} |c_k| < A_0^2 |c_{i_0}|^2.$$

It is easy to see that with probability 1

$$\lim_{n\to\infty} Q_T(\lambda_{i_0}\omega_0) = A_0 |c_{i_0}|^2.$$

Let $E = \{e\}$ be a space of elementary events and

$$\Psi = \left\{e : \varlimsup_{n\to\infty} \sup_{\omega\in\Phi_\delta} Q_T(\omega) < \lim_{n\to\infty} Q_T(\lambda_{i_0}\omega_0) \text{ for any } 0 < \delta < \infty\right\},$$

$$\Phi_\delta = \left\{\omega : \omega \geq 0,\ \ |\lambda_{i_0}\omega_0 - \omega| \geq \delta\right\}.$$

By virtue of previous statements $P(\Psi) = 1$. Suppose now that $\overline{\omega}_T \not\to \omega_0$ with probability 1. Let $\Phi_1 = \{e : \overline{\omega}_T \not\to \omega_0\}$ and an elementary event $e \in \Psi_1 \bigcap \Psi$. Then there exists such a subsequence $T_k(e) \to \infty$, $k \to \infty$ that

$$\overline{\omega}_{T_k(e)}(e) \to \overline{\omega}'(e) \neq \omega_0, \quad 0 \leq \overline{\omega}'(e) \leq \infty.$$

Take

$$0 < \delta(e) < \min\left(\, |\overline{\omega}'(e) - \lambda_{i_0}\omega_0| \, , \, \frac{\Delta\omega_0}{2} \right), \quad \omega'(e) = \lambda_{i_0}\overline{\omega}'(e).$$

Because of the inequality

$$\varlimsup_{n\to\infty} \sup_{\omega\in\Phi_{\delta(e)}} Q^{(e)}_{T_k(e)}(\omega) < \lim_{n\to\infty} Q^{(e)}_{T_k(e)}(\lambda_{i_0}\omega_0)$$

we have

$$\varlimsup_{n\to\infty} Q^{(e)}_{T_k(e)}\left(\omega^{(e)}_{T_k(e)}\right) < \lim_{n\to\infty} Q^{(e)}_{T_k(e)}(\lambda_{i_0}\omega_0).$$

From the other side by definition

$$Q^{(e)}_{T_k(e)}\left(\omega^{(e)}_{T_k(e)}\right) \geq Q^{(e)}_{T_k(e)}(\lambda_{i_0}\omega_0),$$

$$\lim_{n\to\infty} Q^{(e)}_{T_k(e)}\left(\omega^{(e)}_{T_k(e)}\right) \geq \lim_{n\to\infty} Q^{(e)}_{T_k(e)}(\lambda_{i_0}\omega_0).$$

We have obtained a contradiction. Consequently, $P(\Psi_1) = 0$ and $\overline{\omega}_T \to \omega_0$, $T \to \infty$ with probability 1. The theorem is proved.

Lemma 33 *Under the conditions of the preceding theorem*

$$\lim_{T\to\infty} Q_T(\omega_T) = \lim_{T\to\infty} Q_T(\lambda_{i_0}\omega_0) = A_0^2 \, |c_{i_0}|^2$$

with probability 1.

Proof. By definition

$$Q_T(\omega_T) \geq Q_T(\lambda_{i_0}\omega_0).$$

That is why

$$0 \leq Q_T(\omega_T) - Q_T(\lambda_{i_0}\omega_0) =$$

$$= I_T(\omega_T) - I_T(\lambda_{i_0}\omega_0) + \left| \frac{A_0}{T} \int_0^T \varphi(\omega_0 t)e^{i\,\omega_T t}\,dt \right|^2 - \left| \frac{A_0}{T} \int_0^T \varphi(\omega_0 t)e^{i\,\lambda_{i_0}\omega_0 t}\,dt \right|^2.$$

Then with probability 1

$$\lim_{T \to \infty} Q_T(\omega_T) = \lim_{T \to \infty} Q_T(\lambda_{i_0}\omega_0) = A_0^2 \, |c_{i_0}|^2.$$

The lemma is proved.

Theorem 44 *Let the conditions of Theorem 43 be fulfilled. Then with probability 1*

$$T\left\{\frac{\omega_T}{\lambda_{i_0}} - \omega_0\right\} \to 0, \quad T \to \infty.$$

Proof. It follows from the proof of Lemma 33 that

$$\left|\frac{1}{T} \int_0^T \varphi(\omega_0 t) e^{i\,\omega_T\,t}\,dt\right|^2 - \left|\frac{1}{T} \int_0^T \varphi(\omega_0 t) e^{i\,\lambda_{i_0}\omega_0\,t}\,dt\right|^2 \to 0 \qquad (4.10)$$

as $T \to \infty$ with probability 1. It can be shown by simple composition that

$$\lim_{T \to \infty}\left|\frac{1}{T} \int_0^T \varphi(\omega_0 t) e^{i\,\omega_T\,t}\,dt\right|^2 = |c_{i_0}|^2 \lim_{T \to \infty} \frac{\sin^2 \dfrac{(\lambda_{i_0}\omega_0 - \omega_T)\,T}{2}}{\left[\dfrac{\lambda_{i_0}\omega_0 - \omega_T}{2}\,T\right]^2}.$$

Taking into consideration (4.10) we obtain that

$$\frac{\sin^2 \dfrac{(\lambda_{i_0}\omega_0 - \omega_T)\,T}{2}}{\left[\dfrac{\lambda_{i_0}\omega_0 - \omega_T}{2}\,T\right]^2} \to 1, \quad T \to \infty$$

with probability 1. Consequently, $T\left(\dfrac{\omega_T}{\lambda_{i_0}} - \omega_0\right) \to 0, T \to \infty$ with probability 1. The theorem is proved.

It is evident that the following assertion follows from Lemma 33.

Theorem 45 *Suppose that conditions A4 − A6 are fulfilled and (4.8) takes place. Then the value*

$$A_T = |c_{i_0}|^{-1} Q_T^{1/2}(\omega_T)$$

is a strongly consistent estimate of A_0.

Let us proceed now to investigation of asymptotical distribution of variables ω_T and A_T. At first we shall prove the following auxiliary statement.

Lemma 34 *Let conditions $A4 - A6$ and (4.8) be satisfied, and $f(\lambda_{i_0}\omega_0) > 0$. Then the following representation takes place:*

$$T^{-1/2}\frac{\partial Q_T(\lambda_{i_0}\omega_0)}{\partial\omega} = \zeta_{T1} + \zeta_{T2}, \tag{4.11}$$

where ζ_{T1} is asymptotically normal random variable with zero mean and variance $\sigma^2 = \frac{1}{3}\pi A_0^2|c_{i_0}|^2 f(\lambda_{i_0}\omega_0)$, $\zeta_{T2} \to 0$ as $T \to \infty$ in probability.

Proof. We have the representation

$$\frac{1}{T^{1/2}}\frac{\partial Q_T(\lambda_{i_0}\omega_0)}{\partial\omega} = \frac{1}{T^{5/2}}\frac{\partial}{\partial\omega}\left\{\left[\int_0^T x(t)\cos\omega t\,dt\right]^2 + \right.$$

$$\left. + \left[\int_0^T x(t)\sin\omega t\,dt\right]^2\right\}_{\omega=\lambda_{i_0}\omega_0} = \frac{2}{T^{5/2}}\left\{-\int_0^T\left[A_0\varphi(\omega_0 t) + n(t)\right]\times\right.$$

$$\times\cos(\lambda_{i_0}\omega_0 t)\,dt\cdot\int_0^T t\left[A_0\varphi(\omega_0 t) + n(t)\right]\sin(\lambda_{i_0}\omega_0 t)\,dt +$$

$$+ \int_0^T\left[A_0\varphi(\omega_0 t) + n(t)\right]\sin(\lambda_{i_0}\omega_0 t)\,dt\int_0^T t\left[A_0\varphi(\omega_0 t) + n(t)\right]\times$$

$$\left. \times\cos(\lambda_{i_0}\omega_0 t)\,dt\right\}. \tag{4.12}$$

Consider each term of the right-hand side of (4.12) separately (4.12). We suppose below that one has convergence in probability. Then,

$$\lim_{T\to\infty}\frac{A_0}{T^{5/2}}\int_0^T\varphi(\omega_0 t)\cos(\lambda_{i_0}\omega_0 t)\,dt\int_0^T t\,n(t)\sin(\lambda_{i_0}\omega_0 t)\,dt =$$

$$= \lim_{T\to\infty} \frac{A_0}{T^{5/2}} \int_0^T \sum_{\nu=-\infty}^\infty c_\nu \, e^{i\,\lambda_\nu \omega_0 t} \cos(\lambda_{i_0}\omega_0)\, dt \int_0^T t\, n(t) \sin(\lambda_{i_0}\omega_0\, t)\, dt =$$

$$= A_0 \frac{c_{i_0} + c_{-i_0}}{2} \lim_{T\to\infty} \int_0^T t\, n(t) \sin(\lambda_{i_0}\omega_0\, t)\, dt, \qquad (4.13)$$

$$\lim_{T\to\infty} \frac{A_0}{T^{5/2}} \int_0^T \varphi(\omega_0\, t) \sin(\lambda_{i_0}\omega_0\, t)\, dt \int_0^T t\, n(t) \cos(\lambda_{i_0}\omega_0\, t)\, dt =$$

$$= A_0 \lim_{T\to\infty} \frac{A_0}{T^{3/2}} \int_0^T t\, n(t) \cos(\lambda_{i_0}\omega_0\, t)\, dt \cdot \frac{c_{-i_0} - c_{i_0}}{2i}, \qquad (4.14)$$

$$\lim_{T\to\infty} \frac{A_0}{T^{5/2}} \int_0^T n(t)\, \cos(\lambda_{i_0}\omega_0\, t)\, dt \int_0^T t \sum_{\nu=-\infty}^\infty c_\nu \, e^{i\,\lambda_\nu \omega_0 t} \sin(\lambda_{i_0}\omega_0)\, dt =$$

$$= A_0 \frac{c_{-i_0} - c_{i_0}}{4i} \lim_{T\to\infty} \frac{A_0}{T^{1/2}} \int_0^T n(t)\, \cos(\lambda_{i_0}\omega_0\, t)\, dt, \qquad (4.15)$$

$$\lim_{T\to\infty} \frac{A_0}{T^{5/2}} \int_0^T n(t)\, \sin(\lambda_{i_0}\omega_0\, t)\, dt \int_0^T t \sum_{\nu=-\infty}^\infty c_\nu \, e^{i\,\lambda_\nu \omega_0 t} \cos(\lambda_{i_0}\omega_0\, t)\, dt =$$

$$= A_0 \frac{c_{i_0} + c_{-i_0}}{4} \lim_{T\to\infty} \frac{A_0}{T^{1/2}} \int_0^T n(t) \sin(\lambda_{i_0}\omega_0\, t)\, dt, \qquad (4.16)$$

$$\lim_{T\to\infty} \left\{ \frac{A_0^2}{T^{1/2}} \left[-\int_0^T \varphi(\omega_0\, t) \cos(\lambda_{i_0}\omega_0\, t)\, dt \int_0^T t\, \varphi(\omega_0\, t) \sin(\lambda_{i_0}\omega_0\, t)\, dt + \right.\right.$$

$$\left.\left. + \int_0^T \varphi(\omega_0\, t) \sin(\lambda_{i_0}\omega_0\, t)\, dt \int_0^T t\, \varphi(\omega_0\, t) \cos(\lambda_{i_0}\omega_0\, t)\, dt \right] \right\} =$$

$$= A_0^2 \left[-\frac{c_{-i_0} + c_{i_0}}{2} \cdot \frac{c_{-i_0} - c_{i_0}}{4i} + \frac{c_{-i_0} - c_{i_0}}{2i} \cdot \frac{c_{-i_0} + c_{i_0}}{4} \right] = 0, \qquad (4.17)$$

$$\lim_{T \to \infty} \frac{A_0}{T^{5/2}} \left[-\int_0^T n(t) \cos(\lambda_{i_0} \omega_0 t) \, dt \int_0^T t\, n(t) \sin(\lambda_{i_0} \omega_0 t) \, dt + \right.$$

$$\left. + \int_0^T n(t) \sin(\lambda_{i_0} \omega_0 t) \, dt \int_0^T t\, n(t) \cos(\lambda_{i_0} \omega_0 t) \, dt \right] = 0 \qquad (4.18)$$

because of Lemma 32.

Then (4.11) follows from (4.12) - (4.18), where

$$\zeta_{T1} = - \left[c_{i_0} + c_{-i_0} \right] \frac{A_0}{T^{3/2}} \int_0^T t\, n(t) \sin(\lambda_{i_0} \omega_0 t) \, dt -$$

$$- \frac{c_{i_0} - c_{-i_0}}{i} \frac{A_0}{T^{3/2}} \int_0^T t\, n(t) \cos(\lambda_{i_0} \omega_0 t) \, dt +$$

$$+ \frac{c_{i_0} - c_{-i_0}}{2i} \frac{A_0}{T^{1/2}} \int_0^T t\, n(t) \cos(\lambda_{i_0} \omega_0 t) \, dt +$$

$$+ \frac{c_{i_0} + c_{-i_0}}{2} \frac{A_0}{T^{1/2}} \int_0^T t\, n(t) \sin(\lambda_{i_0} \omega_0 t) \, dt.$$

Let

$$a_{i_0} = \frac{c_{i_0} + c_{-i_0}}{2}, \quad b_{i_0} = \frac{c_{i_0} - c_{-i_0}}{2i},$$

$$a_T(t) = \frac{A_0}{\sqrt{T}} \left[b_{i_0} \cos(\lambda_{i_0} \omega_0 t) \left(1 - \frac{2t}{T} \right) + a_{i_0} \sin(\lambda_{i_0} \omega_0 t) \left(1 - \frac{2t}{T} \right) \right]. \qquad (4.19)$$

Then the function $\rho(u)$ responding to the function $a_T(t)$ as defined in Theorem 42 has the form

$$\rho(u) = \lim_{T \to \infty} \frac{A_0^2}{T W^2(T)} \int_0^T \left(1 - \frac{2(t + |u|)}{T} \right) \left(1 - \frac{2t}{T} \right) \left\{ a_{i_0} \sin\left(\lambda_{i_0} \omega_0 (t + |u|) \right) + \right.$$

$+b_{i_0} \cos\left(\lambda_{i_0}\omega_0\left(t+|u|\right)\right)\right\} \left\{a_{i_0}\sin(\lambda_{i_0}\omega_0\,t)+b_{i_0}\cos(\lambda_{i_0}\omega_0\,t)\right\} dt = \cos(\lambda_{i_0}\omega_0\,u).$

Consequently, the variable

$$\zeta_{T1} = \int\limits_0^T a_T(t)\,n(t)\,dt,$$

where $a_T(t)$ is defined by (4.19), is asymptotically normal with mean 0 and variance

$$\sigma^2 = \frac{A_0^2}{6}\,2\pi\,|c_{i_0}|^2\,f(\lambda_{i_0}\omega_0) = \frac{\pi}{3}\,A_0^2\,|c_{i_0}|^2\,f(\lambda_{i_0}\omega_0).$$

The lemma is proved.

Lemma 35 *Suppose that conditions $A4-A6$ and (4.8) are fulfilled. Then for any random variable $\breve{\omega}_T$, satisfying with probability 1 the inequality*

$$|\breve{\omega}_T - \lambda_{i_0}\omega_0| \le |\omega_T - \lambda_{i_0}\omega_0|$$

for all $T > 0$, we have

$$\frac{1}{T^2}\,\frac{\partial^2 Q_T(\breve{\omega}_T)}{\partial\omega^2} \to -\frac{1}{6}\,|c_{i_0}|^2, \quad T \to \infty$$

in probability.

Proof. The following relationship is valid:

$$\frac{1}{T^2}\,\frac{\partial^2 Q_T(\breve{\omega}_T)}{\partial\omega^2} =$$

$$= \frac{2}{T^4}\left\{\left[\int\limits_0^T t\,x(t)\,\sin(\breve{\omega}_T\,t)\,dt\right]^2 + \left[\int\limits_0^T t\,x(t)\,\cos(\breve{\omega}_T\,t)\,dt\right]^2 - \right.$$

$$- \int\limits_0^T x(t)\,\cos(\breve{\omega}_T\,t)\,dt \int\limits_0^T t^2\,x(t)\,\cos(\breve{\omega}_T\,t)\,dt -$$

$$\left. - \int\limits_0^T x(t)\,\sin(\breve{\omega}_T\,t)\,dt \int\limits_0^T t^2\,x(t)\,\sin(\breve{\omega}_T\,t)\,dt \right\}.$$

It is easy to see that because of Theorem 44 and Lemma 32

$$\frac{2}{T^4} \left[\int_0^T t\, x(t)\, \sin(\breve{\omega}_T\, t)\, dt \right]^2 \to \frac{A_0^2}{2}\, b_{i_0}^2,$$

$$\frac{2}{T^4} \left[\int_0^T t\, x(t)\, \cos(\breve{\omega}_T\, t)\, dt \right]^2 \to \frac{A_0^2}{2}\, a_{i_0}^2,$$

$$\frac{2}{T^4} \int_0^T x(t)\, \cos(\breve{\omega}_T\, t)\, dt \int_0^T t^2\, x(t)\, \cos(\breve{\omega}_T\, t)\, dt \to \frac{2}{3}\, A_0^2\, a_{i_0}^2,$$

$$\frac{2}{T^4} \int_0^T x(t)\, \sin(\breve{\omega}_T\, t)\, dt \int_0^T t^2\, x(t)\, \sin(\breve{\omega}_T\, t)\, dt \to \frac{2}{3}\, A_0^2\, b_{i_0}^2$$

in probability. Then

$$\lim_{T \to \infty} \frac{1}{T^2} \frac{\partial^2 Q_T(\breve{\omega}_T)}{\partial \omega^2} = -A_0^2 \left(\frac{2}{3} - \frac{1}{3} \right) (a_{i_0}^2 + b_{i_0}^2) = \frac{|c_{i_0}|^2}{6}\, A_0^2\, b_{i_0}$$

in probability. The lemma is proved.

Theorem 46 *Assume that conditions A1−A3 and (4.8) are satisfied, $f(\lambda_{i_0}\omega_0) >$ 0. Then the variable $T^{3/2}(\omega_T - \lambda_{i_0}\omega_0)$ is asymptotically normal with zero mean and variance*

$$\sigma^2 = 12\pi\, A_0^{-2}\, |c_{i_0}|^{-2}\, f(\lambda_{i_0}\omega_0).$$

Proof. Because $\dfrac{\omega_T}{\lambda_{i_0}} \to \omega_0$ as $T \to \infty$ with probability 1, we have that with probability converging to 1 as $T \to \infty$, ω_T is an internal point of $[0, \infty)$. With the same probability $Q_T'(\omega_T) = 0$ and

$$Q_T'(\lambda_{i_0}\omega_0) + Q_T''(\breve{\omega}_T)\, (\omega_T - \lambda_{i_0}\omega_0) = 0, \tag{4.20}$$

where $'$ and $''$ mean the first and second derivatives respectively, $\breve{\omega}_T$ is some random variable satisfying with probability 1 the inequality

$$|\breve{\omega}_T - \lambda_{i_0}\omega_0| \le |\omega_T - \lambda_{i_0}\omega_0|, \quad T > 0.$$

It follows from (4.20) that

$$\omega_T - \lambda_{i_0}\omega_0 = -\frac{Q'_T(\lambda_{i_0}\omega_0)}{Q''_T(\breve{\omega}_T)}. \tag{4.21}$$

The equality (4.21) is equivalent to the following one:

$$T^{3/2}(\omega_T - \lambda_{i_0}\omega_0) = -\frac{T^{-1/2}Q'_T(\lambda_{i_0}\omega_0)}{T^{-2}Q''_T(\breve{\omega}_T)}. \tag{4.22}$$

The denominator of the right-hand side of (4.22) converges in probability to the value

$$A = -\frac{|c_{i_0}|^2}{6}A_0^2.$$

That is why taking into consideration Lemmas 34 and 35 we obtain the assertion of the theorem. The proof is complete.

We proceed in finding the asymptotical distribution of the variable $\sqrt{T}\,(A_T - A_0)$.

Theorem 47 *Let conditions* $A4 - A6$ *and (4.8) be fulfilled,* $f(\lambda_{i_0}\omega_0) > 0$. *Then the variable* $\xi_T = \sqrt{T}\,(A_T - A_0)$ *is asymptotically normal with parameters* $\left(0, \pi\,|c_{i_0}|^{-2}\,f(\lambda_{i_0}\omega_0)\right)$.

Proof. We have

$$\xi_T = \sqrt{T}\,\left[\,|c_{i_0}|^{-1}Q_T^{1/2}(\omega_T) - A_0\right] = \sqrt{T}\,\left[\,|c_{i_0}|^{-2}Q_T(\omega_T) - A_0^2\right] \times$$

$$\times\left[\,|c_{i_0}|^{-1}Q_T^{1/2}(\omega_T) + A_0\right]^{-1}.$$

It follows from Theorem 45 that

$$|c_{i_0}|^{-1}Q_T^{1/2}(\omega_T) + A_0 \to 2\,A_0, \quad T \to \infty$$

with probability 1. Let us check that

$$\sqrt{T}\,[Q_T(\omega_T) - Q_T(\lambda_{i_0}\omega_0)] \to 0, \quad T \to \infty$$

in probability. In fact

$$\sqrt{T}\left[Q_T(\omega_T) - Q_T(\lambda_{i_0}\omega_0)\right] = \sqrt{T}\,Q_T'(\lambda_{i_0}\omega_0)\,(\omega_T - \lambda_{i_0}\omega_0) +$$

$$+ \frac{1}{2}\sqrt{T}\,Q_T''(\breve{\omega}_T)\,(\omega_T - \lambda_{i_0}\omega_0)^2$$

with some value $\breve{\omega}_T$ such that

$$|\breve{\omega}_T - \lambda_{i_0}\omega_0| \le |\omega_T - \lambda_{i_0}\omega_0|$$

with probability 1 as $T \to \infty$. The variable

$$\sqrt{T}\,Q_T'(\lambda_{i_0}\omega_0)\,(\omega_T - \lambda_{i_0}\omega_0) = \frac{1}{\sqrt{T}}\,Q_T'(\lambda_{i_0}\omega_0)\,T\,(\omega_T - \lambda_{i_0}\omega_0)$$

converges in probability to 0 as $T \to \infty$ according to Lemma 34 and Theorem 44. The variable

$$\frac{1}{2}\sqrt{T}\,Q_T''(\breve{\omega}_T)\,(\omega_T - \lambda_{i_0}\omega_0)^2 = \frac{1}{2T^2}\sqrt{T}\,Q_T''(\breve{\omega}_T)\,T^{3/2}\,(\omega_T - \lambda_{i_0}\omega_0) \times$$

$$\times\,(\omega_T - \lambda_{i_0}\omega_0)$$

converges in probability to 0 because of Lemma 35, Theorems 44 and 46. From this it follows that the asymptotical distribution of ξ_T coincide with the asymptotical distribution of the variable

$$\beta_T = \frac{1}{2A_0}\sqrt{T}\left[\,|c_{i_0}|^{-2}Q_T(\lambda_{i_0}\omega_0) - A_0^2\right].$$

The following relationship is valid for β_T in the sense of convergence in probability:

$$\lim_{T\to\infty}\frac{\sqrt{T}}{2A_0}\left[\,|c_{i_0}|^{-2}\,Q_T(\lambda_{i_0}\omega_0) - A_0^2\right] =$$

$$= \lim_{T\to\infty}\frac{\sqrt{T}}{2A_0}\left[\,|c_{i_0}|^{-2}\left|\frac{1}{T}\int_0^T\left[A_0\,\varphi(\omega_0\,t) + n(t)\right]e^{i\lambda_{i_0}\omega_0\,t}\,dt\right|^2 - A_0^2\right] =$$

$$= \lim_{T\to\infty}\frac{\sqrt{T}}{2A_0}\left[\,|c_{i_0}|^{-2}\frac{1}{T^2}\int_0^T\left[A_0\,\varphi(\omega_0\,t) + n(t)\right]e^{i\lambda_{i_0}\omega_0\,t}\,dt\times$$

$$\times \int_0^T \left[A_0 \, \varphi(\omega_0 \, t) + n(t)\right] e^{-i \, \lambda_{i_0} \omega_0 \, t} \, dt - A_0^2\right] = \lim_{T \to \infty} \frac{1}{2} \, |c_{i_0}|^{-2} \times$$

$$\times \left\{c_{-i_0} \frac{1}{\sqrt{T}} \int_0^T n(t) \, e^{-i \, \lambda_{i_0} \omega_0 \, t} \, dt + c_{i_0} \frac{1}{\sqrt{T}} \int_0^T n(t) \, e^{i \, \lambda_{i_0} \omega_0 \, t} \, dt\right\} =$$

$$= \frac{1}{2|c_{i_0}|^2} \lim_{T \to \infty} \left\{\frac{1}{\sqrt{T}} \int_0^T n(t) \left[c_{-i_0} \, e^{-i \, \lambda_{i_0} \omega_0 \, t} + c_{i_0} \, e^{i \, \lambda_{i_0} \omega_0 \, t}\right] =$$

$$= \lim_{T \to \infty} \frac{1}{|c_{i_0}|^2} \frac{1}{\sqrt{T}} \int_0^T n(t) \left[a_{i_0} \, \cos(\lambda_{i_0} \omega_0 \, t) - b_{i_0} \, \sin(\lambda_{i_0} \omega_0 \, t)\right] \, dt.$$

Denote

$$a_T(t) = \frac{1}{|c_{i_0}|^2} \frac{1}{\sqrt{T}} \left[a_{i_0} \, \cos(\lambda_{i_0} \omega_0 \, t) - b_{i_0} \, \sin(\lambda_{i_0} \omega_0 \, t)\right].$$

Then

$$\rho(u) = \lim_{T \to \infty} \frac{1}{W^2(T)} \frac{1}{|c_{i_0}|^4 \, T} \int_0^T \left\{a_{i_0} \, \cos[\lambda_{i_0} \omega_0 (t + |u|)] - \right.$$

$$\left. -b_{i_0} \, \sin\left[\lambda_{i_0} \omega_0 \, (t + |u|)\right]\right\} \left\{a_{i_0} \, \cos(\lambda_{i_0} \omega_0 \, t) - b_{i_0} \, \sin(\lambda_{i_0} \omega_0 \, t)\right\} \, dt = \cos(\lambda_{i_0} \omega_0 \, u).$$

We obtain that the variable β_T and consequently ξ_T are asymptotically normal with the parameters $\left(0, \pi \, |c_{i_0}|^{-2} \, f(\lambda_{i_0} \omega_0)\right)$. The theorem is proved.

4.3   ASYMPTOTIC BEHAVIOR OF PERIODOGRAM ESTIMATES OF THE SECOND TYPE

Suppose that the unknown parameter is $\omega_0 \in (\underline{\omega}, \overline{\omega})$, $\underline{\omega} > 0$, $\overline{\omega} < \infty$. Let us consider the functional

$$\tilde{Q}_T(\omega) = \left|\frac{2}{T} \int_0^T x(t) \, \varphi(\omega \, t) \, dt\right|^2$$

and we select as the estimate of ω_0 that value $\tilde{\omega}_T \in [\underline{\omega}, \overline{\omega}]$ for which $\tilde{Q}_T(\omega)$ reaches the maximal value. Because $\tilde{Q}_T(\omega)$ is continuous function of ω with probability 1, the estimate $\tilde{\omega}_T$ is determined with probability 1 and by virtue of Theorem 6 similarly to Lemma 30 it can be chosen to be a random variable. Let us prove an assertion concerning the strong consistency of $\tilde{\omega}_T$.

Theorem 48 *Let conditions A4 − A6 be satisfiesd. Then $\tilde{\omega}_T \to \omega$, $T \to \infty$ with probability 1.*

Proof. We fix ω and consider the behavior of the quantity $\tilde{Q}_T(\omega)$ as $T \to \infty$:

$$\tilde{Q}_T(\omega) = \left| \frac{1}{T} \int_0^T x(t)\,\varphi(\omega\,t)\,dt \right|^2 = \left| \frac{1}{T} \int_0^T [\,A_0\,\varphi(\omega_0\,t) + n(t)\,]\,\varphi(\omega\,t)\,dt \right|^2 =$$

$$= \left| \frac{1}{T} \int_0^T A_0\,\varphi(\omega_0\,t)\,\varphi(\omega\,t)\,dt \right|^2 + \tilde{I}_T(\omega),$$

$$\tilde{I}_T(\omega) = \frac{1}{T^2} \left| \int_0^T n(t)\,\varphi(\omega\,t)\,dt \right|^2 + 2\frac{A_0}{T^2} \int_0^T \varphi(\omega_0\,t)\,\varphi(\omega\,t)\,dt \int_0^T n(t)\,\varphi(\omega\,t)\,dt.$$

Because of condition $A6$

$$\frac{1}{T} \left| \int_0^T \varphi(\omega_0\,t)\,\varphi(\omega\,t)\,dt \right| \le c, \quad 0 < c < \infty.$$

That is why using Lemma 31 we have

$$\sup_\omega \tilde{I}_T(\omega) \to 0, \quad T \to \infty$$

with probability 1.

Let

$$\tilde{\psi}_T(\omega_0, \omega) = \frac{A_0}{T} \int_0^T \varphi(\omega_0\,t)\,\varphi(\omega\,t)\,dt.$$

The following equality is valid for $\widetilde{\psi}_T(\omega_0, \omega)$:

$$\widetilde{\psi}_T(\omega_0, \omega) = \sum_{j,k=-\infty}^{\infty} \frac{A_0}{T} c_j c_k \int_0^T e^{i(\lambda_j \omega_0 + \lambda_k \omega)t} \, dt.$$

Let $0 < \widetilde{\delta} < \dfrac{\Delta\omega}{2}$. Suppose that for some j and k $|\lambda_j \omega_0 - \lambda_k \omega| \le \widetilde{\delta}$. Then for any $l \ne j$

$$|\lambda_l \omega_0 - \lambda_k \omega| = \left|(\lambda_j \omega_0 - \lambda_k \omega) + (\lambda_l - \lambda_j)\omega_0\right| \ge \frac{\Delta\omega}{2} > \widetilde{\delta}. \qquad (4.23)$$

Analogously, for any $l \ne k$

$$|\lambda_j \omega_0 - \lambda_l \omega| = \left|(\lambda_j \omega_0 - \lambda_k \omega) - (\lambda_l - \lambda_k)\omega\right| \ge \frac{\Delta\omega}{2} > \widetilde{\delta}. \qquad (4.24)$$

Let $|\lambda_1 \omega_0 - \lambda_k \omega| \le \widetilde{\delta}$. We show that for every $l \ne 1$

$$\left|\lambda_l \omega_0 - \lambda_1 \omega\right| \ge \widetilde{\delta}. \qquad (4.25)$$

Inequality (4.25) is obvious for $l \le 0$. Suppose that $l > 0$. Then

$$|\lambda_l \omega_0 - \lambda_1 \omega| = \left|(\lambda_1 \omega_0 - \lambda_k \omega) + (\lambda_l - \lambda_1)\omega_0 + (\lambda_k - \lambda_1)\omega\right| \ge \widetilde{\delta}.$$

Analogously, if $|\lambda_k \omega_0 - \lambda_1 \omega| \le \widetilde{\delta}$ then for any integer $l \ne 1$

$$\left|\lambda_1 \omega_0 - \lambda_l \omega\right| \ge \widetilde{\delta}.$$

Let us also note that

$$|\lambda_k \omega| \ge \lambda_1 \underline{\omega} > 2\widetilde{\delta}, \quad |\lambda_j \omega_0| \ge \lambda_1 \underline{\omega} > 2\widetilde{\delta}, \quad k, j \ne 0. \qquad (4.26)$$

Then taking into account (4.23) - (4.26) for any $\delta > 0$ and $0 < \widetilde{\delta} < \dfrac{\Delta\omega}{2}$ we have with probability 1

$$\overline{\lim_{T\to\infty}} \sup_{\omega \in [\underline{\omega}, \overline{\omega}], |\omega - \omega_0| \ge \delta} \widetilde{Q}_T(\omega) \le \overline{\lim_{T\to\infty}} \sup_{\omega \in [\underline{\omega}, \overline{\omega}], |\omega - \omega_0| \ge \min\left(\delta, \frac{\delta}{\lambda_1}\right)} \left|\widetilde{\psi}_T(\omega_0, \omega)\right|^2 =$$

$$= \overline{\lim_{T\to\infty}} \sup_{\omega \in [\underline{\omega}, \overline{\omega}], |\omega - \omega_0| \ge \min\left(\delta, \frac{\widetilde{\delta}}{\lambda_1}\right)} \left|\frac{A_0}{T} \sum_{j,k=-\infty}^{\infty} c_j c_k \int_0^T e^{i(\lambda_j \omega_0 + \lambda_k \omega)t} \, dt\right|^2 \le$$

$$\leq A_0^2 \varlimsup_{T\to\infty} \sup_{\omega\in[\underline{\omega},\overline{\omega}],|\omega-\omega_0|\geq\min\left(\delta,\frac{\widetilde{\delta}}{\lambda_1}\right)} \left| |c_0|^2 + \frac{1}{T}\sum_{j,k=1}^{\infty} c_j\,\overline{c}_k \int_0^T e^{i(\lambda_j\omega_0-\lambda_k\omega)\,t}\,dt + \right.$$

$$\left. +\frac{1}{T}\sum_{j,k=1}^{\infty} \overline{c}_j\,c_k \int_0^T e^{-(\lambda_j\omega_0-\lambda_k\omega)\,t}\,dt \right|^2 \leq A_0^2 \sup_{\omega\in[\underline{\omega},\overline{\omega}],|\omega-\omega_0|\geq\min\left(\delta,\frac{\widetilde{\delta}}{\lambda_1}\right)} \left[|c_0|^2 + \right.$$

$$\left. +2\sum_{j,k=1}^{\infty} |c_j\,c_k|\,\delta_{jk}(\omega_0,\omega) \right]^2 ,$$

where

$$\delta_{jk}(\omega_0,\omega) = \begin{cases} 1, & |\lambda_j\omega_0 - \lambda_k\omega| < \min(\lambda_1\delta,\widetilde{\delta}) \\ 0, & |\lambda_j\omega_0 - \lambda_k\omega| \geq \min(\lambda_1\delta,\widetilde{\delta}) \end{cases}, \quad j,k \geq 1.$$

By virtue of (4.23) - (4.26) the function $\delta_{jk}(\omega_0,\omega)$ possesses the following properties:

a) if $\delta_{j_0 k_0}(\omega_0,\omega) = 1$ then $\delta_{j_0 k}(\omega_0,\omega) = \delta_{j k_0}(\omega_0,\omega) = 0$, $k \neq k_0$, $j \neq j_0$;

b) if $\delta_{1k}(\omega_0,\omega) = 1$ then $\delta_{l_1 k}(\omega_0,\omega) = 0$ for $l \neq 1$;

c) if $\delta_{j1}(\omega_0,\omega) = 1$ then $\delta_{1l}(\omega_0,\omega) = 0$ for $l \neq 1$.

Therefore

$$\varlimsup_{T\to\infty} \sup_{\omega\in[\underline{\omega},\overline{\omega}],|\omega-\omega_0|\geq\delta} \widetilde{Q}_T(\omega) \leq A_0^2 \left[|c_0|^2 + |c_1|^2 + 2\sum_{j=2}^{\infty} |c_j|^2 \right]^2 =$$

$$= A_0^2 \left[\sum_{j=-\infty}^{\infty} |c_j|^2 - |c_1|^2 \right]^2 < A_0^2 \left[\sum_{j=-\infty}^{\infty} |c_j|^2 \right]^2 .$$

It is also easy to see that with probability 1

$$\lim_{T\to\infty} \widetilde{Q}_T(\omega) = A_0^2 \left[\sum_{j=-\infty}^{\infty} |c_j|^2 \right]^2 .$$

Consequently, with probability 1 the following inequality is satisfied:

$$\varlimsup_{T \to \infty} \sup_{\omega \in [\underline{\omega}, \overline{\omega}], |\omega - \omega_0| \geq \delta} \tilde{Q}_T(\omega) < \lim_{T \to \infty} \tilde{Q}_T(\omega_0).$$

The rest of the proof is similar to Theorem 43. The proof is complete.

The following assertion will be written without proof. It is analogous to proof of Theorem 44.

Theorem 49 *Let the conditions of Theorem 48 be satisfied. Then with probability 1*

$$T(\tilde{\omega}_T - \omega) \to 0, \quad T \to \infty.$$

Lemma 36 *Suppose that the conditions of Theorem 48 are fulfilled, and*

$$\sum_{j=-\infty}^{\infty} |\lambda_j c_j| < \infty, \quad \sum_{j=-\infty}^{\infty} \lambda_j^2 |c_j|^2 f(\lambda_j \omega_0) > 0.$$

Then the following representation holds

$$\frac{1}{T^{1/2}} \frac{\partial \tilde{Q}_T(\omega_0)}{\partial \omega} = \xi_{T1} + \xi_{T2}, \tag{4.27}$$

where ξ_{T1} is an asymptotically normal random variable with zero mean and variance

$$\sigma^2 = \frac{8}{3} \pi A_0^2 \left[\sum_{\nu=-\infty}^{\infty} |c_\nu|^2 \right]^2 \sum_{\nu=-\infty}^{\infty} \lambda_\nu^2 |c_\nu|^2 f(\lambda_\nu \omega), \tag{4.28}$$

and $\xi_{T2} \to 0$, $T \to \infty$ in probability.

Proof. The representation

$$\frac{1}{T^{1/2}} \frac{\partial \tilde{Q}_T(\omega_0)}{\partial \omega} = \frac{2}{T^{5/2}} \int_0^T [A_0 \varphi(\omega_0 t) + n(t)] \varphi(\omega_0 t) \, dt \times$$

$$\times \int_0^T t [A_0 \varphi(\omega_0 t) + n(t)] \varphi'(\omega_0 t) \, dt \tag{4.29}$$

is valid. We consider each term in the right-hand side of (4.29) separately. Denote

$$I_T^1 = \frac{2}{T} \int_0^T x(t)\,\varphi(\omega_0 t)\,dt \; \frac{2}{T^{3/2}} \int_0^T t\,\varphi(\omega_0 t)\,\varphi'(\omega_0 t)\,dt. \qquad (4.30)$$

For the second factor in the right-hand side of (4.30) we have

$$\lim_{T\to\infty} \frac{2}{T^{3/2}} \int_0^T t \sum_{j=-\infty}^{\infty} c_k e^{i\lambda_k \omega_0 t} \sum_{j=-\infty}^{\infty} i\lambda_l c_l e^{i\lambda_i \omega_0 t}\,dt =$$

$$= \lim_{T\to\infty} \frac{2}{T^{3/2}} \int_0^T t \sum_{k,l=-\infty}^{\infty} i\lambda_l c_k c_l e^{i(\lambda_k+\lambda_l)\omega_0 t}\,dt =$$

$$= \lim_{T\to\infty} \frac{2}{T^{3/2}} \left[\int_0^T t \sum_{k\neq -l} i\lambda_l c_k c_l e^{i(\lambda_k+\lambda_l)\omega_0 t}\,dt + \right.$$

$$\left. + i \int_0^T t\,dt \sum_{k=-\infty}^{\infty} \lambda_k c_k c_{-k} \right] = 0.$$

Then taking into account Theorem 42 we obtain that $\lim_{T\to\infty} I_T^1 = 0$ in probability.

It also follows from Theorem 42 that

$$\lim_{T\to\infty} \frac{1}{T^{5/2}} \int_0^T \varphi(\omega_0 t)\,n(t)\,dt \int_0^T t\,\varphi'(\omega_0 t)\,n(t)\,dt = 0$$

in probability.

Let us consider the behavior of the quantity

$$I_T^2 = \frac{2}{T^{5/2}} \int_0^T \varphi^2(\omega_0 t)\,dt \int_0^T t\,\varphi'(\omega_0 t)\,n(t)\,dt$$

as $T \to \infty$. Obviously

$$\lim_{T \to \infty} \frac{2}{T} \int_0^T \varphi^2(\omega_0 t) \, dt = 2 \sum_{\nu=-\infty}^{\infty} |c_\nu|^2.$$

That is why

$$\lim_{T \to \infty} I_T^2 = 2 \sum_{\nu=-\infty}^{\infty} |c_\nu|^2 \lim_{T \to \infty} \frac{1}{T^{3/2}} \int_0^T t \, \varphi'(\omega_0 t) \, n(t) \, dt$$

in probability. Let

$$a_T(t) = A_0 \, \frac{2 \sum\limits_{\nu=-\infty}^{\infty} |c_\nu|^2}{T^{3/2}} \, t \, \varphi'(\omega_0 t).$$

Then the function $\rho(u)$ defined in Theorem 42 is

$$\rho(u) = \frac{1}{\sum\limits_{\nu=-\infty}^{\infty} \lambda_\nu^2 |c_\nu|^2} \sum_{\nu=-\infty}^{\infty} \lambda_\nu^2 |c_\nu|^2 \, e^{-i \lambda_\nu \, \omega_0 |u|}.$$

Consequently, the limit distribution of the quantity I_T^2 as $T \to \infty$ is normal with mean 0 and variance

$$\sigma^2 = \frac{8}{3} \pi A_0^2 \left[\sum_{\nu=-\infty}^{\infty} |c_\nu|^2 \right]^2 \sum_{\nu=-\infty}^{\infty} \lambda_\nu^2 |c_\nu|^2 \, f(\lambda_\nu \, \omega_0).$$

The lemma is proved.

Lemma 37 *Let the conditions of Lemma 36 be satisfied. Then for any random variable $\overline{\omega}_T$ satisfying the inequality $|\overline{\omega}_T - \omega_0| \le |\tilde{\omega}_T - \omega_0|$ with probability 1, the following equality takes place:*

$$\lim_{T \to \infty} \frac{1}{T^2} \frac{\partial^2 \widetilde{Q}_T(\overline{\omega}_T)}{\partial \omega^2} = -\frac{2}{3} A_0^2 \sum_{\nu=-\infty}^{\infty} |c_\nu|^2 \sum_{\nu=-\infty}^{\infty} \lambda_\nu^2 |c_\nu|^2 \tag{4.31}$$

in probability.

Proof. The following equality is valid:

$$\frac{1}{T^2}\frac{\partial^2 Q_T(\overline{\omega}_T)}{\partial \omega^2} = \frac{2}{T^4}\left\{\left[\int_0^T t\,x(t)\,\varphi'(\overline{\omega}_T\,t)\,dt\right]^2 + \right.$$

$$\left. + \int_0^T x(t)\,\varphi(\overline{\omega}_T\,t)\,dt\int_0^T t^2\,x(t)\,\varphi''(\overline{\omega}_T\,t)\,dt\right\}.$$

Using Theorem 49 and Lemma 31 it is easy to see that in probability

$$\lim_{T\to\infty}\frac{1}{T^2}\int_0^T t\,x(t)\,\varphi'(\overline{\omega}_T\,t)\,dt = 0,$$

$$\lim_{T\to\infty}\frac{1}{T^3}\int_0^T t^2\,n(t)\,\varphi''(\overline{\omega}_T\,t)\,dt = 0.$$

Therefore, in probability

$$\lim_{T\to\infty}\left(\frac{1}{T^2}\frac{\partial^2 \widetilde{Q}_T(\overline{\omega}_T)}{\partial \omega^2}\right) = \lim_{T\to\infty}\left[\frac{2\,A_0^2}{T^4}\int_0^T \varphi(\omega_0\,t)\,\varphi(\overline{\omega}_T\,t)\,dt\times\right.$$

$$\left. \times\int_0^T t^2\,\varphi(\omega_0\,t)\,\varphi''(\overline{\omega}_T\,t)\,dt\right] = -\frac{2}{3}\,A_0^2\sum_{\nu=-\infty}^{\infty}|c_\nu|^2\sum_{\nu=-\infty}^{\infty}\lambda_\nu^2\,|c_\nu|^2.$$

The lemma is proved.

Theorem 50 *Assume that the conditions of Lemma 37 are valid. Then the quantity $T^{3/2}(\widetilde{\omega}_T - \omega_0)$ is asymptotically normal with zero mean and variance*

$$\sigma^2 = 6\,\pi\,A_0^{-2}\left[\sum_{\nu=-\infty}^{\infty}|c_\nu|^2\right]^{-2}\sum_{\nu=-\infty}^{\infty}\lambda_\nu^2\,|c_\nu|^2\,f(\lambda_\nu\,\omega_0).$$

Proof. Since $\widetilde{\omega}_T \to \omega_0$, $T \to \infty$ with probability 1, we have that with probability tending to 1 as $T \to \infty$ $\widetilde{\omega}_T \in (\underline{\omega}_T, \overline{\omega}_T)$. Hence with the same probability

$\tilde{Q}'_T(\tilde{\omega}_T) = 0$ and the equality

$$\tilde{Q}'_T(\omega_0) + \tilde{Q}''_T(\overline{\omega}_T)(\overline{\omega}_T - \omega_0) = 0 \tag{4.32}$$

holds with some random variable $\overline{\omega}_T$ satisfying with probability 1 the inequality $|\overline{\omega}_T - \omega_0| \le |\tilde{\omega}_T - \omega_0|$, $T > 0$.

It follows from (4.32) that

$$\tilde{\omega}_T - \omega_0 = -\frac{\tilde{Q}'_T(\omega_0)}{\tilde{Q}''_T(\overline{\omega}_T)}. \tag{4.33}$$

The equality (4.33) is equivalent to the following one:

$$T^{3/2}(\tilde{\omega}_T - \omega_0) = -\frac{T^{-1/2}\,\tilde{Q}'_T(\omega_0)}{T^{-2}\,\tilde{Q}''_T(\overline{\omega}_T)}. \tag{4.34}$$

The denominator in the right-hand side of (4.34) tends in probability to the quantity

$$-\frac{2}{3}A_0^2 \sum_{\nu=-\infty}^{\infty} |c_\nu|^2 \sum_{\nu=-\infty}^{\infty} \lambda_\nu^2 |c_\nu|^2.$$

Now taking into account (4.27) and (4.28), we obtain the assertion of the theorem.

We stop briefly on the estimation of A_0. Let us take the following quantity as an estimate of A_0:

$$\tilde{A}_T = \left(\sum_{\nu=-\infty}^{\infty} |c_\nu|^2 \right)^{-1} \tilde{Q}_T^{1/2}(\tilde{\omega}_T).$$

The following assertion can be proved similarly to the estimates of the first type.

Theorem 51 *Let the conditions of Theorem 48 be satisfied. Then* $\tilde{A}_T \to A_0$, $T \to \infty$ *with probability 1.*

Theorem 52 *Suppose that the conditions of Lemma 37 are fulfilled. Then the quantity* $\sqrt{T}\,(\tilde{A}_T - A_0)$ *is asymptotically normal with mean 0 and variance*

$$\sigma^2 = 2\pi \sum_{k=-\infty}^{\infty} |c_k|^2 f(\lambda_k\,\omega_0) \left(\sum_{k=-\infty}^{\infty} |c_k|^2 \right)^{-2}.$$

4.4 PERIODOGRAM ESTIMATES IN $\mathcal{R}^M$

We will briefly consider the case when the random field $\left\{ n(\vec{t}),\ \vec{t} \in \mathcal{R}^m \right\}$ satisfies conditions $A1, A2$, and function $\varphi(\vec{t}),\ \vec{t} \in \mathcal{R}^m$ satisfies condition $A3$.

Let us consider the functional

$$Q_T(\vec{\omega}) = \left| \frac{1}{T^m} \int\limits_{[0,T]^m} x(\vec{t})\, e^{i\,(\vec{\omega},\,\vec{t})}\, dt \right|^2 ,$$

where $x(\vec{t})$ is defined by (4.6). Suppose that $\vec{\omega}_T = (\omega_{T1}, \ldots, \omega_{Tm})$ is the value of the parameter $\vec{\omega} = (\omega_1, \ldots, \omega_{1n})$, $\omega_i \geq 0$, for which the functional reaches its maximal value. Obviously, the quantity $\vec{\omega}_T$ is defined with probability 1 and because of Theorem 6 it can be chosen to be a random variable. The following assertion is valid.

Theorem 53 *Assume that conditions $A1 - A3$ and*

$$|c_{k_1,\ldots,k_m}| > |c_{l_1,\ldots,l_m}|, \quad (l_1,\ldots,l_m) \neq (k_1,\ldots,k_m),$$

$$k_i > 0, \quad l_i \geq 0, \quad i = \overline{1,m} \qquad (4.35)$$

are satisfied. Then

$$\frac{\omega_{Ti}}{\lambda_{k_i}} \to \omega_{0i}, \quad i = \overline{1,m}, \quad as \quad T \to \infty$$

with probability 1.

The proof of the theorem is absolutely analogous to the proof of Theorem 43.

Analogously to Theorem 44 we can prove stronger assertion.

Theorem 54 *Under the conditions of Theorem 52*

$$T\left(\frac{\omega_{Ti}}{\lambda_{k_i}} - \omega_{0i} \right) \to 0, \quad i = \overline{1,m},$$

with probability 1 as $T \to \infty$.

We shall consider periodogram estimates of the second type for multidimensional parameter. Let

$$\widetilde{Q}_T(\vec{\omega}) = \left| \frac{1}{T^m} \int_0^T x(\vec{t})\, \varphi(\vec{\omega} \cdot \vec{t})\, dt \right|^2,$$

where $x(\vec{t})$ is defined by (4.6). Suppose that unknown m-dimension parameter $\vec{\omega}_0$ belongs to the domain

$$\Omega = \left\{ \vec{\omega} = (\omega_1, \ldots, \omega_m), \quad 0 < \underline{\omega}_i < \omega_i < \overline{\omega}_i < \infty, \quad i = \overline{1, m} \right\}.$$

We select $\widetilde{\vec{\omega}}_T \in \overline{\Omega} = \left\{ \vec{\omega} = (\omega_1, \ldots, \omega_m),\, 0 < \underline{\omega}_i < \omega_i < \overline{\omega}_i < \infty,\, i = \overline{1, m} \right\}$, for which the functional $\widetilde{Q}_T(\vec{\omega})$ reaches the maximal value, as an estimate of $\vec{\omega}_0$. As above it can be shown that $\widetilde{\vec{\omega}}_T$ exists and it is a random variable.

The following assertion is proved analogously to Theorem 48.

Theorem 55 *Let conditions $A1 - A3$ be fulfilled. Then $\widetilde{\vec{\omega}}_T \to \vec{\omega}_0$, $T \to \infty$ with probability 1.*

As for 1-dimension case stronger assertion takes place.

Theorem 56 *Under the conditions of Theorem 55*

$$T\left(\widetilde{\vec{\omega}}_T - \vec{\omega}_0 \right) \to 0, \quad T \to \infty$$

with probability 1.

Now we shall consider estimates of A_0. Let

$$A_T = |c_{k_1, \ldots, k_m}|^{-1} Q_T^{1/2}(\vec{\omega}_T),$$

$$\widetilde{A}_T = \left[\sum_{k_1, \ldots, k_m = -\infty}^{\infty} |c_{k_1, \ldots, k_m}|^2 \right]^{-1} \widetilde{Q}_T^{1/2}(\widetilde{\vec{\omega}}_T).$$

The following assertions are valid.

Theorem 57 *Suppose that conditions $A1 - A3$ and (4.35) are satisfied. Then the quantity A_T is strongly consistent estimate of A_0.*

Theorem 58 *Let conditions $A1 - A3$ be fulfilled. Then the quantity $\widetilde{A}_T$ is strongly consistent estimate of A_0.*

5

NONPARAMETRIC
IDENTIFICATION PROBLEMS

In this chapter the nonparametric optimization and estimation models are considered.

The empirical estimates for the general model of stochastic optimization are studied, then the obtained results are applied to the proof of assertions on the strong consistency and asymptotic normality for some nonparametric regression models.

5.1 THE INVESTIGATION OF THE GENERAL PROBLEM

At first we will formulate the auxiliary statements, and then we shall go to the basic problems.

For arbitrary vectors

$$\vec{i} = (i_j)_{j=1}^m, \quad \vec{n} = (n_j)_{j=1}^m \in \mathcal{Z}_+^m, \quad m \geq 1,$$

we will write $\vec{i} \leq \vec{n}$ for the case when $i_j \leq n_j$, $j = \overline{1, m}$. If for some vector $\vec{n} = (n_j)_{j=1}^m \in \mathcal{N}^m$ we have $n_j > q$, $j = \overline{1, m}$, where $q \in \mathcal{N}$, then we write $\vec{n} > q$.

We will need the following assertion

199

Theorem 59 [15]. Let $(\Omega, \mathcal{F}, P)$ be a probabilistic space, $\{\mathcal{F}_n, n \geq 1\}$ be a sequence of σ-algebras such that $\mathcal{F}_n \subset \mathcal{F}_{n+1}$, $\mathcal{F}_n \subset \mathcal{F}$, $n \geq 1$. Suppose that K is a compact subset of some Banach space with a norm $\|\cdot\|$, and

$$\Big\{ Q_n(s) = Q_n(s, \omega), \quad (s, \omega) \in K \times \Omega, \quad n \geq 1 \Big\}$$

is a sequence of real functions, satisfying to the following conditions:

1) for fixed n and each $s \in K$ the function $Q_n(s, \omega) \,:\, \Omega \to \mathcal{R}$ is $\mathcal{F}_n$-measurable;

2) for fixed n, ω the function $Q_n(s, \omega) \,:\, K \to \mathcal{R}$ is continuous on K;

3) for some $s_0 \in K$ and every $s \in K$

$$\lim_{n \to \infty} P\Big\{ \big| Q_n(s) - \Phi(s; s_0) \big| > \varepsilon \Big\} = 0, \quad \varepsilon > 0$$

with some real function $\Phi(s; s_0)$, $s \in K$, continuous on K and such that

$$\Phi(s; s_0) > \Phi(s_0; s_0), \quad s \neq s_0;$$

4) there exist $\gamma_0 > 0$ and a function $c(\gamma) \,:\, \mathcal{R}_+ \to \mathcal{R}$, $c(\gamma) \to 0$, $\gamma \to 0$, such that for any $s' \in K$ and $0 < \gamma < \gamma_0$

$$\lim_{n \to \infty} P\left\{ \sup_{s\,:\,\|s-s'\|<\gamma} |Q_n(s) - Q_n(s')| < c(\gamma) \right\} = 1.$$

Denote

$$s_n = \arg\min_{s \in K} Q_n(s).$$

Then the sequence $\{s_n, n \geq 1\}$ converges in probability to s_0:

for any $\varepsilon > 0$

$$\lim_{n \to \infty} P\Big\{ \big\| s_n - s_0) \big\| > \varepsilon \Big\} = 0,$$

and the sequence $\{Q_n(s_n), n \geq 1\}$ converges in probability to $\Phi(s_0; s_0)$:

for each $\varepsilon > 0$

$$\lim_{n \to \infty} P\Big\{ \big| Q_n(s) - \Phi(s_0; s_0) \big| > \varepsilon \Big\} = 0.$$

It is easy to see that Theorem 59 is valid for the multidimensional parameter $\vec{n} \in \mathcal{N}^m$, $m \geq 1$ [15].

Lemma 38 *Let* $\left\{ \eta(\vec{i},\vec{n}), \ \vec{i} \in \mathbb{Z}_+^m, \ \vec{i} \leq \vec{n}, \ \vec{n} \in \mathcal{N}^m \right\}$, $m \geq 1$, *be the family of real random values, defined on the probabilistic space* $(\Omega, \mathcal{G}, P)$, *and suppose that for any* $\vec{n} \in \mathcal{N}^m$ *the random variables* $\eta(\vec{i},\vec{n})$, $\vec{i} \leq \vec{n}$ *are independent. If there exists such a constant* α *that for all* $\vec{n} \in \mathcal{N}^m$, $\vec{i} \leq \vec{n}$ *we have* $E\left(\eta(\vec{i},\vec{n})\right)^2 \leq \alpha$, *then* $S_{\vec{n}} - E\,S_{\vec{n}} \to 0$, $\vec{n} \to \infty$, *in probability, where*

$$S_{\vec{n}} = \frac{1}{\prod\limits_{j=1}^{m} n_j} \sum_{\vec{i} \leq \vec{n}} \eta(\vec{i},\vec{n}).$$

If for all $\vec{n} \in \mathcal{N}^m$, $\vec{i} \leq \vec{n}$ *we have* $E\left(\eta(\vec{i},\vec{n})\right)^4 \leq \alpha$, *then with probability 1* $S_{\vec{n}} - E\,S_{\vec{n}} \to 0$, $\vec{n} \to \infty$.

Proof. For any $\vec{n} \in \mathcal{N}^m$ we have because of the independence of values $\eta(\vec{i},\vec{n})$, $\vec{i} \leq \vec{n}$

$$E\left(S_{\vec{n}} - E\,S_{\vec{n}}\right)^2 = E\left(\frac{1}{\prod\limits_{j=1}^{m} n_j} \sum_{\vec{i} \leq \vec{n}} \left(\eta(\vec{i},\vec{n}) - E\,\eta(\vec{i},\vec{n})\right)\right)^2 =$$

$$= \frac{1}{\prod\limits_{j=1}^{m} (n_j)^2} \sum_{\vec{i} \leq \vec{n}} E\left(\eta(\vec{i},\vec{n}) - E\,\eta(\vec{i},\vec{n})\right)^2 \leq \frac{\alpha_1 \prod\limits_{j=1}^{m} (n_j+1)}{\prod\limits_{j=1}^{m} (n_j)^2} \leq \frac{\alpha_2}{\prod\limits_{j=1}^{m} n_j};$$

$$E\left(S_{\vec{n}} - E\,S_{\vec{n}}\right)^4 = E\left(\frac{1}{\prod\limits_{j=1}^{m} n_j} \sum_{\vec{i} \leq \vec{n}} \left(\eta(\vec{i},\vec{n}) - E\,\eta(\vec{i},\vec{n})\right)\right)^4 =$$

$$= \frac{1}{\prod\limits_{j=1}^{m} (n_j)^4} \left(\sum_{\vec{i} \leq \vec{n}} E\left(\eta(\vec{i},\vec{n}) - E\eta(\vec{i},\vec{n}) \right)^4 + \right.$$

$$\left. +3 \sum_{\substack{\vec{i},\vec{k} \leq \vec{n} \\ \vec{i} \neq \vec{k}}} E\left(\eta(\vec{i},\vec{n}) - E\eta(\vec{i},\vec{n}) \right)^2 E\left(\eta(\vec{k},\vec{n}) - E\eta(\vec{k},\vec{n}) \right)^2 \right) \leq$$

$$\leq \frac{\alpha_3 \prod\limits_{j=1}^{m} (n_j + 1) + 3\alpha_3 \prod\limits_{j=1}^{m} (n_j + 1)^2}{\prod\limits_{j=1}^{m} (n_j)^4} \leq \frac{\alpha_4}{\prod\limits_{j=1}^{m} (n_j)^2},$$

where α_1, α_2, α_3, α_4 are constants.

Now Chebishev inequality and Borel-Cantelli Lemma imply the proof of Lemma 38.

Let $\left\{ \xi(\vec{t}) = \xi(\vec{t}, \omega), \ \vec{t} \in [0,1]^l \right\}$, $l \geq 1$ be the random field, defined on the complete probabilistic space $(\Omega, \mathcal{G}, P)$, with values in some metric space $(Y, \mathcal{B}(Y))$ and traectories of the field $\xi(\vec{t})$ are continuous with probability 1.

Denote by $C[0,1]^k$, $k \geq 1$ a space of real functions, defined and continuous on $[0,1]^k$, with the norm

$$\|\alpha\| = \max_{\vec{t} \in [0,1]^k} |\alpha(\vec{t})|, \quad \alpha = \alpha(\vec{t}) \in C[0,1]^k.$$

Suppose that K is a compact subset in $C[0,1]^{m+l}$, $m \geq 1$. Then for some c we have $\|\alpha\| \leq c$, $\alpha \in K$. Denote $I = [-c, c]$.

The continuous function $f : [0,1]^m \times [0,1]^l \times I \times Y \to \mathcal{R}$ is known, and

$$\sup_{\vec{s} \in [0,1]^m, \ \vec{t} \in [0,1]^l} E \left| f\left(\vec{s}, \vec{t}, \alpha(\vec{s}, \vec{t}), \xi(\vec{t}) \right) \right| < \infty, \quad \alpha \in K.$$

There are observations

$$\left\{\xi(\vec{i}, \vec{n}, \vec{t}\,), \quad \vec{i} \in \mathcal{Z}_+^m, \quad \vec{i} \leq \vec{n}, \quad \vec{t} \in [0,1]^l\right\}, \quad \vec{n} \in \mathcal{N}^m,$$

and the following conditions below are fulfilled:

1) for all $\vec{i} \leq \vec{n}$ $\left\{\xi(\vec{i}, \vec{n}, \vec{t}\,) = \xi(\vec{i}, \vec{n}, \vec{t}, \omega), \vec{t} \in [0,1]^l\right\}$ is the random field, defined on the space $(\Omega, \mathcal{G}, P)$, with values in $\left(Y, \mathcal{B}(Y)\right)$, and its finite dimension distributions coincide with those of $\xi(\vec{t}\,)$;

2) σ-algebras $\mathcal{F}(\vec{i}, \vec{n})$, $\vec{i} \leq \vec{n}$ are independent, where

$$\mathcal{F}(\vec{i}, \vec{n}) = \sigma\left\{\xi(\vec{i}, \vec{n}, \vec{t}\,), \vec{t} \in [0,1]^l\right\};$$

3) for each $\vec{i} \leq \vec{n}$ traectories of the field $\xi(\vec{i}, \vec{n}, \vec{t}\,)$ are continuous with probability 1.

We need to find points of minimum and a minimal value of the functional

$$F(\alpha) = E \int_{[0,1]^{m+l}} f\left(\vec{s}, \vec{t}, \alpha(\vec{s}, \vec{t}\,), \xi(\vec{t}\,)\right) d\vec{s}\, d\vec{t}, \quad \alpha \in K. \tag{5.1}$$

This problem is approximated by minimization of the functional

$$F_{\vec{n}}(\alpha) = F_{\vec{n}}(\alpha, \omega) = \frac{1}{\prod\limits_{j=1}^m n_j} \sum_{\vec{i} \leq \vec{n}} \times$$

$$\times \int_{[0,1]^l} f\left(\vec{s}\,(\vec{i}, \vec{n}), \vec{t}, \alpha(\vec{s}\,(\vec{i}, \vec{n}), \vec{t}\,), \xi(\vec{i}, \vec{n}, \vec{t}\,)\right) d\vec{t}, \tag{5.2}$$

where $\alpha \in K$, $\vec{n} = (n_j)_{j=1}^m$, $\vec{i} = (i_j)_{j=1}^m$, $\vec{s}\,(\vec{i}, \vec{n}) = \left(\dfrac{i_j}{n_j}\right)_{j=1}^m$.

It follows from continuity of the function f and Lebesgue theorem of limit transition that for any $\vec{n} \in \mathcal{N}^m$ and $\omega \in \Omega'$, $P(\Omega') = 1$, the function $F_{\vec{n}}(\alpha, \omega)$,

$\alpha \in K$ is continuous and has at least one minimum point. For each $\vec{n} \in \mathcal{N}^m$ and $\alpha \in K$ the mapping $F_{\vec{n}}(\alpha, \omega)$, $\omega \in \Omega'$ is $\mathcal{G}'_{\vec{n}}$-measurable, where

$$\mathcal{G}'_{\vec{n}} = \mathcal{G}_{\vec{n}} \bigcap \Omega', \quad \mathcal{G}_{\vec{n}} = \sigma \left\{ \xi \left(\vec{i}, \vec{n}, \vec{t} \right), \quad \vec{t} \in [0,1]^l, \quad \vec{i} \leq \vec{n} \right\}.$$

Consequently by virtue of Theorem 6 for every $\vec{n}$ the point of minimum for function (5.2) can be chosen $\mathcal{G}'_{\vec{n}}$-measurable as a function of ω.

Theorem 60 *Let the following conditions be satisfied:*

1) *there exists such a constant α that for all $\vec{s} \in [0,1]^m$, $\vec{t} \in [0,1]^l$*

$$E \left\{ \max_{\alpha \in K} \left(f \left(\vec{s}, \vec{t}, \alpha(\vec{s}, \vec{t}), \xi(\vec{t}) \right) \right)^2 \right\} \leq \alpha;$$

2) *there exists a single point of minimum α_0 of the function (5.1).*

Then

$$\|\alpha_{\vec{n}} - \alpha_0\| \to 0, \quad F_{\vec{n}}(\alpha_{\vec{n}}) \to F(\alpha_0), \quad \vec{n} \to \infty,$$

in probability, where $\alpha_{\vec{n}} = \alpha_{\vec{n}}(\omega)$, $\vec{n} \in \mathcal{N}^m$, $\omega \in \Omega'$ is a minimum point of the function (5.2), $\mathcal{G}'_{\vec{n}}$-measurable in ω.

If in condition 1) the inequality is

$$E \left\{ \max_{\alpha \in K} \left(f \left(\vec{s}, \vec{t}, \alpha(\vec{s}, \vec{t}), \xi(\vec{t}) \right) \right)^4 \right\} \leq \alpha;$$

then

$$P \left\{ \|\alpha_{\vec{n}} - \alpha_0\| \to 0, \quad F_{\vec{n}}(\alpha_{\vec{n}}) \to F(\alpha_0), \quad \vec{n} \to \infty \right\} = 1.$$

Proof. Let us check conditions of Theorem 59 for the family of functions

$$\left\{ F_{\vec{n}} : K \times \Omega' \to \mathcal{R}, \quad \vec{n} \in \mathcal{N}^m \right\}.$$

It is evident that conditions 1) and 2) are fulfilled.

Fix $\alpha \in K$. Fubini theorem and condition 1) of Theorem 60 imply that

$$E F_{\vec{n}}(\alpha) \to F(\alpha), \quad \vec{n} \to \infty. \tag{5.3}$$

Denote

$$\eta(\vec{i}, \vec{n}) = \int_{[0,1]^l} f\left(\vec{s}\,(\vec{i}, \vec{n}), \vec{t}, \alpha(\vec{s}\,(\vec{i}, \vec{n}), \vec{t}), \xi(\vec{i}, \vec{n}, \vec{t})\right) d\vec{t},$$

$$\vec{i} \leq \vec{n}, \quad \vec{n} \in \mathcal{N}^m.$$

For any $\vec{i}, \vec{n}$ the mapping $\eta(\vec{i}, \vec{n}, \omega)$, $\omega \in \Omega'$ is $\tilde{\mathcal{F}}(\vec{i}, \vec{n})$-measurable, where $\tilde{\mathcal{F}}(\vec{i}, \vec{n}) = \left\{ A : A = B \bigcap \Omega', B \in \mathcal{F}(\vec{i}, \vec{n}) \right\}$. For each $\vec{n}$ the random variables $\eta(\vec{i}, \vec{n})$, $\vec{i} \leq \vec{n}$ are independent. By virtue of Cauchy-Buniakowski inequality, Fubini theorem and condition 1)

$$E\left(\eta(\vec{i}, \vec{n})\right)^2 \leq \alpha, \quad \vec{i} \leq \vec{n}, \quad \vec{n} \in \mathcal{N}^m.$$

Then we can apply Lemma 38 to the variables $\eta(\vec{i}, \vec{n})$ and obtain that

$$F_{\vec{n}}(\alpha) - E F_{\vec{n}}(\alpha) \to 0, \quad \vec{n} \to \infty$$

in probability. Now (5.3) implies

$$P\left\{ \left| F_{\vec{n}}(\alpha) - F(\alpha) \right| > \varepsilon \right\} \to 0, \quad \vec{n} \to \infty, \quad \varepsilon > 0. \tag{5.4}$$

Then the validity of condition 3) of Theorem 59 follows from the continuity of the function (5.1) and condition 2) of Theorem 60.

Denote

$$\psi(\gamma, \vec{s}, \vec{t}, y) = \sup_{\alpha, \tilde{\alpha} \in K \,:\, \|\alpha - \tilde{\alpha}\| < \gamma} \left| f\left(\vec{s}, \vec{t}, \alpha(\vec{s}, \vec{t}), y\right) - f\left(\vec{s}, \vec{t}, \tilde{\alpha}(\vec{s}, \vec{t}), y\right) \right|,$$

$$\gamma > 0, \quad \vec{s} \in [0,1]^m, \quad \vec{t} \in [0,1]^l, \quad y \in Y.$$

For all $\gamma > 0$ the mapping $\psi(\gamma, \vec{s}, \vec{t}, y)$, $(\vec{s}, \vec{t}, y) \in [0,1]^m \times [0,1]^l \times Y$ is continuous. For any $\bar{\alpha} \in K$, $\gamma > 0$, $\vec{n} \in \mathcal{N}^m$ and $\omega \in \Omega'$

$$\sup_{\alpha \in K \,:\, \|\alpha - \bar{\alpha}\| < \gamma} \left| F_{\vec{n}}(\alpha) - F_{\vec{n}}(\bar{\alpha}) \right| \leq \frac{1}{\prod_{j=1}^m n_j} \sum_{\vec{i} \leq \vec{n}} \times$$

$$\times \int_{[0,1]^l} \psi\left(\gamma, \vec{s}\,(\vec{i},\vec{n}),\, \vec{t},\, \xi(\vec{i},\vec{n},\vec{t})\right)\, d\,\vec{t} = \zeta_{\vec{n}}\,(\gamma). \qquad (5.5)$$

We can prove similarly to (5.4) that for each $\gamma > 0$

$$P\left\{\left|\zeta_{\vec{n}}\,(\gamma) - E\int_{[0,1]^l} \psi\left(\gamma,\vec{s},\,\vec{t},\,\xi(\vec{t})\right)\, d\,\vec{s}\; d\,\vec{t}\right| > \varepsilon\right\} \to 0, \quad \vec{n} \to \infty,\ \varepsilon > 0.$$
$$(5.6)$$

Denote

$$c(\gamma) = E\int_{[0,1]^l} \psi\left(\gamma,\vec{s},\,\vec{t},\,\xi(\vec{t})\right)\, d\,\vec{s}\; d\,\vec{t} + \gamma, \quad \gamma > 0.$$

We have $c(\gamma) \to 0$, $\gamma \to 0$. That is why from (5.5), (5.6) we obtain the condition 4) of Theorem 59.

Then all conditions of Theorem 59 are satisfied, and the first part of Theorem 60 is proved.

The second part of the theorem is proved similarly to the first one by using Theorem 7. The proof is complete.

Let us consider the following problem.

Suppose that K is some compact subset of $C\,[0,1]^m$. Then there exists such c that $\|\alpha\| \le c$, $\alpha \in K$. Denote $I = [-c, c]$.

Assume that $\xi = \xi(\omega)$ is a random element defined on a complete probabilistic space $(\Omega, \mathcal{G}, P)$, with values in some measurable space $(Y, \mathcal{U})$. We know the function $f : [0,1]^m \times I \times Y \to \mathcal{R}$, satisfying the conditions:

1) for every $y \in Y$ the mapping $f(\vec{t}, x, y)$, $\vec{t} \in [0,1]^m$, $x \in I$ is continuous;

2) for all $\vec{t} \in [0,1]^m$, $x \in I$ the mapping $f(\vec{t}, x, y)$, $y \in Y$ is $\mathcal{U}$-measurable;

3) for each function $\alpha \in K$ there exists such a constant α that

$$E\left|f\left(\vec{t},\, \alpha(\vec{t}),\, \xi\right)\right| \le \alpha, \quad \vec{t} \in [0,1]^m.$$

Let us have independent observations $\left\{ \xi(\vec{i}, \vec{n}), \ \vec{i} \in \mathbb{Z}_+^m, \ \vec{i} \leq \vec{n} \right\}$ of the random element ξ; $\vec{n} \in \mathcal{N}^m$.

The problem is to find the function from K which is the point of minimum for the functional

$$F(\alpha) = E \int_{[0,1]^l} f\left(\vec{t}, \alpha(\vec{t}), \xi\right) d\vec{t}, \quad \alpha \in K, \tag{5.7}$$

and the minimal value of this functional. We approximate this problem by minimization of a functional

$$F_{\vec{n}}(\alpha) = F_{\vec{n}}(\alpha, \omega) =$$

$$= \frac{1}{\prod\limits_{j=1}^{m} n_j} \sum_{\vec{i} \leq \vec{n}} f\left(\vec{t}\left(\vec{i}, \vec{n}\right), \alpha\left(\vec{t}\left(\vec{i}, \vec{n}\right)\right), \xi(\vec{i}, \vec{n})\right), \quad \alpha \in K, \tag{5.8}$$

where $\vec{n} = (n_j)_{j=1}^m$, $\vec{i} = (i_j)_{j=1}^m$, $\vec{t}\left(\vec{i}, \vec{n}\right) = \left(\dfrac{i_j}{n_j}\right)_{j=1}^m$.

Fix an arbitrary vector $\vec{n} \in \mathcal{N}^m$. For every $\omega \in \Omega$ the function $F_{\vec{n}}(\alpha, \omega)$, $\alpha \in K$ is continuous on K and consequently has at least one minimum point. For each $\alpha \in K$ the mapping $F_{\vec{n}}(\alpha, \omega)$, $\omega \in \Omega$ is $\mathcal{G}_{\vec{n}}$-measurable, where

$$\mathcal{G}_{\vec{n}} = \sigma \left\{ \xi\left(\vec{i}, \vec{n}\right), \ \vec{i} \leq \vec{n} \right\}$$

That is why the minimum point for function (5.8) can be chosen $\mathcal{G}_{\vec{n}}$-measurable.

Theorem 61 *Suppose that the conditions below are fulfilled:*

1) there exists such a constant α that for all $\vec{t} \in [0,1]^m$

$$E \left\{ \max_{\alpha \in K} \left(f\left(\vec{t}, \alpha(\vec{t}), \xi\right) \right)^2 \right\} \leq \alpha;$$

2) the function (5.7) has the single minimum point α_0.

Then

$$\|\alpha_{\vec{n}} - \alpha_0\| \to 0, \quad F_{\vec{n}}(\alpha_{\vec{n}}) \to F(\alpha_0), \quad \vec{n} \to \infty,$$

in probability, where $\alpha_{\vec{n}} = \alpha_{\vec{n}}(\omega)$, $\vec{n} \in \mathcal{N}^m$, $\omega \in \Omega$ is a minimum point of the function (5.8), which is $\mathcal{G}_{\vec{n}}$-measurable in ω.

If in condition 1) the inequality is

$$E \left\{ \max_{\alpha \in K} \left(f \left(\vec{t}, \alpha(\vec{t}), \xi \right) \right)^4 \right\} \leq \alpha;$$

then

$$P \left\{ \|\alpha_{\vec{n}} - \alpha_0\| \to 0, \quad F_{\vec{n}}(\alpha_{\vec{n}}) \to F(\alpha_0), \quad \vec{n} \to \infty \right\} = 1.$$

Proof. For proving of the first part of the theorem it is sufficient to check conditions of Theorem 59 for the family of functions $\left\{ F_{\vec{n}} : K \times \Omega \to \mathcal{R}, \vec{n} \in \mathcal{N}^m \right\}$.

Conditions 1) and 2) are evidently fulfilled.

Then fix a function $\alpha \in K$. By virtue of condition 1) of the theorem

$$E F_{\vec{n}}(\alpha) \to F(\alpha), \quad \vec{n} \to \infty.$$

Denote

$$\eta(\vec{i}, \vec{n}) = f \left(\vec{t} \, (\vec{i}, \vec{n}), \alpha \left(\vec{t} \, (\vec{i}, \vec{n}) \right), \xi(\vec{i}, \vec{n}) \right), \quad \vec{i} \leq \vec{n}, \quad \vec{n} \in \mathcal{N}^m.$$

Because of condition 1) of the theorem and Lemma 38

$$F_{\vec{n}}(\alpha) - E F_{\vec{n}}(\alpha) \to 0, \quad \vec{n} \to \infty$$

in probability. Hence $F_{\vec{n}}(\alpha) \to F(\alpha), \vec{n} \to \infty$ in probability.

Condition 1) of the theorem implies that the functional $F(\alpha)$ is continuous on K. Then condition 3) of Theorem 59 is satisfied.

Denote

$$\psi(\gamma, \vec{t}, y) = \sup_{\alpha, \widetilde{\alpha} \in K \, : \, \|\alpha - \widetilde{\alpha}\| < \gamma} \left| f \left(\vec{t}, \alpha(\vec{t}), y \right) - f \left(\vec{t}, \widetilde{\alpha}(\vec{t}), y \right) \right|,$$

$$\gamma > 0, \quad \vec{t} \in [0, 1]^m, \quad y \in Y.$$

For any $\overline{\alpha}$, $\gamma > 0$, $\vec{n} \in \mathcal{N}^m$ and $\omega \in \Omega$

$$\sup_{\alpha \in K \,:\, \|\alpha - \overline{\alpha}\| < \gamma} \left| F_{\vec{n}}(\alpha) - F_{\vec{n}}(\overline{\alpha}) \right| \le$$

$$\le \frac{1}{\prod\limits_{j=1}^{m} n_j} \sum_{\vec{i} \le \vec{n}} \psi\left(\gamma, \vec{t}\,(\vec{i}, \vec{n}), \xi(\vec{i}, \vec{n})\right) = \zeta_{\vec{n}}(\gamma). \tag{5.9}$$

The properties of continuous mapping imply that for all $\gamma > 0$ and $y \in Y$ the function $\psi(\gamma, \vec{t}, y)$, $\vec{t} \in [0, 1]^m$ is continuous. Then because of condition 1) of the theorem

$$E\zeta_{\vec{n}}(\gamma) \to E \int\limits_{[0,1]^l} \psi(\gamma, \vec{t}, \xi)\, d\,\vec{t}, \quad \vec{n} \to \infty, \quad \gamma > 0.$$

Lemma 38 implies that $\zeta_{\vec{n}}(\gamma) - E\zeta_{\vec{n}}(\gamma) \to 0$, $\vec{n} \to \infty$ in probability, $\gamma > 0$, hence for every $\gamma > 0$

$$\zeta_{\vec{n}}(\gamma) \to E \int\limits_{[0,1]^l} \psi(\gamma, \vec{t}, \xi)\, d\,\vec{t}, \quad \vec{n} \to \infty \tag{5.10}$$

in probability. Denote

$$c(\gamma) = E \int\limits_{[0,1]^l} \psi(\gamma, \vec{t}, \xi)\, d\,\vec{t} + \gamma, \quad \gamma > 0.$$

By virtue of B.Levi theorem of limit transition

$$c(\gamma) \to 0, \quad \gamma \to 0.$$

Then it follows from (5.9),(5.10) that condition 4) of Theorem 59 is fulfilled.

Now we have the first part of Theorem 61. The second part of this theorem follows from Theorem 7. Theorem 61 is proved.

5.2 THE NONPARAMETRIC REGRESSION MODEL WITH OBSERVATIONS IN A FINITE NUMBER OF CURVES ON THE PLANE

Now we will consider some regression models, to which Theorem 60 and 61 are applied.

Let K be a compact subset of $C[0, 1]^{m+l}$, $m, l \geq 1$. Our problem is to estimate the unknown function $\alpha_0 \in K$ by observations

$$x\left(\vec{i}, \vec{n}, \vec{t}\right) = \alpha_0 \left(\vec{s}\left(\vec{i}, \vec{n}\right), \vec{t}\right) + \xi(\vec{i}, \vec{n}, \vec{t}), \quad \vec{t} \in [0, 1]^l, \quad \vec{i} \in \mathcal{Z}_+^m, \quad \vec{i} \leq \vec{n},$$

where

$$\vec{n} = (n_j)_{j=1}^m \in \mathcal{N}^m, \quad \vec{i} = (i_j)_{j=1}^m, \quad \vec{s}\left(\vec{i}, \vec{n}\right) = \left(\frac{i_j}{n_j}\right)_{j=1}^m,$$

$$\left\{\xi(\vec{i}, \vec{n}, \vec{t}), \quad \vec{t} \in [0, 1]^l, \quad \vec{i} \leq \vec{n}\right\}$$

is the family of real random fields defined on a complete probabilistic space $(\Omega, \mathcal{G}, P)$ and satisfying the conditions:

1) σ-algebras $\mathcal{F}(\vec{i}, \vec{n}) = \sigma\left\{\xi(\vec{i}, \vec{n}, \vec{t}), \vec{t} \in [0, 1]^l\right\}$, $\vec{i} \leq \vec{n}$ are independent;

2) for all $\vec{i} \leq \vec{n}$ traectories of the field $\xi(\vec{i}, \vec{n}, \vec{t})$ are continuous with probability 1;

3) the finite-dimension distributions of the random field $\xi(\vec{i}, \vec{n}, \vec{t})$ do not depend on $\vec{i}, \vec{n}$;

4) for every $\vec{t} \in [0, 1]^l$ the zero point is the single median of distribution of the random variables $\xi(\vec{i}, \vec{n}, \vec{t})$, $\vec{i} \leq \vec{n}$, $\vec{n} \in \mathcal{N}^m$;

5) there exists such a constant α that $E\left(\xi(\vec{i}, \vec{n}, \vec{t})\right)^2 \leq \alpha$, $\vec{i} \leq \vec{n}$, $\vec{n} \in \mathcal{N}^m$, $\vec{t} \in [0, 1]^l$.

Let us consider the least modules estimate of the function α_0, which is a minimum point of the functional

$$F_{\vec{n}}(\alpha) = \frac{1}{\prod_{j=1}^{m} n_j} \sum_{\vec{i} \leq \vec{n}} \int_{[0,1]^l} \left| x(\vec{i}, \vec{n}, \vec{t}) - \alpha\left(\vec{s}\,(\vec{i}, \vec{n}), \vec{t}\right) \right| d\,\vec{t} =$$

$$= \frac{1}{\prod_{j=1}^{m} n_j} \sum_{\vec{i} \leq \vec{n}} \int_{[0,1]^l} \left| \alpha_0\left(\vec{s}\,(\vec{i}, \vec{n}), \vec{t}\right) - \alpha\left(\vec{s}\,(\vec{i}, \vec{n}), \vec{t}\right) + \xi(\vec{i}, \vec{n}, \vec{t}) \right| d\,\vec{t},$$

$$(5.11)$$

where $\alpha \in K$.

Theorem 62 *For any* $\vec{n} \in \mathcal{N}^m$ *and* $\omega \in \Omega'$, $P(\Omega') = 1$ *there exists at least one minimum point* $\alpha_{\vec{n}} = \alpha_{\vec{n}}(\omega)$ *for the function (5.11), and for any* $\vec{n}$ *the function* $\alpha_{\vec{n}}(\omega)$, $\omega \in \Omega'$ *can be chosen* $\mathcal{G}'_{\vec{n}}$*-measurable, where*

$$\mathcal{G}'_{\vec{n}} = \mathcal{G}_{\vec{n}} \bigcap \Omega', \quad \mathcal{G}_{\vec{n}} = \sigma\left\{ \xi\left(\vec{i}, \vec{n}, \vec{t}\right), \quad \vec{t} \in [0,1]^l, \quad \vec{i} \leq \vec{n} \right\}.$$

In this case

$$\|\alpha_{\vec{n}} - \alpha_0\| \to 0, \quad F_{\vec{n}}(\alpha_{\vec{n}}) \to E \int_{[0,1]^l} \left| \xi(\vec{0}, \vec{1}, \vec{t}) \right| d\,\vec{t}, \quad \vec{n} \to \infty,$$

in probability.

If in condition 5) the inequality is

$$E\left(\xi(\vec{i}, \vec{n}, \vec{t})\right)^4 \leq \alpha, \quad \vec{i} \leq \vec{n}, \quad \vec{n} \in \mathcal{N}^m, \quad \vec{t} \in [0,1]^l$$

then

$$P\left\{ \|\alpha_{\vec{n}} - \alpha_0\| \to 0, \quad F_{\vec{n}}(\alpha_{\vec{n}}) \to E \int_{[0,1]^l} \left| \xi(\vec{0}, \vec{1}, \vec{t}) \right| d\,\vec{t}, \quad \vec{n} \to \infty \right\} = 1.$$

Proof. We will show that the function α_0 is a single minimum point of the functional

$$F(\alpha) = E \int_{[0,1]^l} \left| \alpha_0\left(\vec{s}, \vec{t}\right) - \alpha(\vec{s}, \vec{t}) + \xi(\vec{0}, \vec{1}, \vec{t}) \right| d\,\vec{s}, d\,\vec{t}, \quad \alpha \in K.$$

Fix $\alpha \in K$, $\alpha \neq \alpha_0$. The continuity of the functions α and α_0 implies existence of some ball S in $\mathcal{R}^{m+l}$, such that $\alpha(\vec{s}, \vec{t}) \neq \alpha_0(\vec{s}, \vec{t})$ for any $(\vec{s}, \vec{t}) \in S$. Then by virtue of Lemma 23

$$E\left|\alpha_0(\vec{s}, \vec{t}) - \alpha(\vec{s}, \vec{t}) + \xi(\vec{0}, \vec{1}, \vec{t})\right| \geq E\left|\xi(\vec{0}, \vec{1}, \vec{t})\right|,$$

$$\vec{s} \in [0,1]^m, \quad \vec{t} \in [0,1]^l,$$

and if $(\vec{s}, \vec{t}) \in S$ then the strict inequality takes the place in the last relation. Fubini theorem implies

$$F(\alpha) - F(\alpha_0) = \int\limits_{[0,1]^{m+l}} \left(E\left|\alpha_0(\vec{s}, \vec{t}) - \alpha(\vec{s}, \vec{t}) + \xi(\vec{0}, \vec{1}, \vec{t})\right| - \right.$$

$$\left. - E\left|\xi(\vec{0}, \vec{1}, \vec{t})\right|\right) d\vec{s}\, d\vec{t} > 0.$$

Now the theorem follows from Theorem 60, because for the function

$$f(\vec{s}, \vec{t}, x, y) = \left|\alpha_0(\vec{s}, \vec{t}) - x + y\right|, \quad \vec{s} \in [0,1]^m,$$

$$t \in [0,1]^l, \quad x \in \mathcal{R}, \quad y \in \mathcal{R}$$

all conditions of Theorem 60 are fulfilled. The proof is complete.

Let us consider now the least squares estimate of the unknown function α_0, which minimizes the functional

$$F_{\vec{n}}(\alpha) = \frac{1}{\prod\limits_{j=1}^{m} n_j} \sum_{\vec{i} \leq \vec{n}} \int\limits_{[0,1]^l} \left(x(\vec{i}, \vec{n}, \vec{t}) - \alpha\left(\vec{s}\,(\vec{i}, \vec{n}), \vec{t}\right)\right)^2 d\vec{t} =$$

$$= \frac{1}{\prod\limits_{j=1}^{m} n_j} \sum_{\vec{i} \leq \vec{n}} \int\limits_{[0,1]^l} \left(\alpha_0\left(\vec{s}\,(\vec{i}, \vec{n}), \vec{t}\right) - \alpha\left(\vec{s}\,(\vec{i}, \vec{n}), \vec{t}\right) + \xi(\vec{i}, \vec{n}, \vec{t})\right)^2 d\vec{t},$$

$$\tag{5.12}$$

$\alpha \in K$.

Theorem 63 *Suppose that the random fields $\xi(\vec{i}, \vec{n}, \vec{t})$ satisfy conditions 1) − 3), 5) and $E\xi(\vec{i}, \vec{n}, \vec{t}) = 0$, $\vec{t} \in [0,1]^l$, $\vec{i} \leq \vec{n}$, $\vec{n} \in \mathcal{N}^m$. Then for all $\vec{n} \in \mathcal{N}^m$,*

$\omega \in \Omega'$, $P(\Omega') = 1$ *the functional* (5.12) *has at least one minimum point* $\alpha_{\vec{n}} = \alpha_{\vec{n}}(\omega)$ *and for any* $\vec{n}$ *it can be chosen* $\mathcal{G}'_{\vec{n}}$*-measurable in* ω,

$$\mathcal{G}'_{\vec{n}} = \mathcal{G}_{\vec{n}} \bigcap \Omega', \quad \mathcal{G}_{\vec{n}} = \sigma\left\{\xi\left(\vec{i}, \vec{n}, \vec{t}\right), \quad \vec{t} \in [0,1]^l, \quad \vec{i} \leq \vec{n}\right\}.$$

In this case

$$\|\alpha_{\vec{n}} - \alpha_0\| \to 0, \quad Q_{\vec{n}}(\alpha_{\vec{n}}) \to 0, \quad \vec{n} \to \infty,$$

in probability, where

$$Q_{\vec{n}}(\alpha_{\vec{n}}) = F_{\vec{n}}(\alpha_{\vec{n}}) - \frac{1}{\prod\limits_{j=1}^{m} n_j} \sum_{\vec{i} \leq \vec{n}} \int_{[0,1]^l} \left(\xi(\vec{i}, \vec{n}, \vec{t})\right)^2 d\vec{t}.$$

If in condition 5) $E\left(\xi(\vec{i}, \vec{n}, \vec{t})\right)^4 \leq \alpha$ *then*

$$P\left\{\|\alpha_{\vec{n}} - \alpha_0\| \to 0, \quad Q_{\vec{n}}(\alpha_{\vec{n}}) \to 0, \quad \vec{n} \to \infty\right\} = 1.$$

Proof. Denote

$$F(\vec{s}, \vec{t}, x, y) = \left(\alpha(\vec{s}, \vec{t}) - x\right)^2 + 2\left(\alpha_0(\vec{s}, \vec{t}) - x\right) y,$$

$$\vec{s} \in [0,1]^m, \quad \vec{t} \in [0,1]^l, \quad \vec{x} \in \mathcal{R}, \quad \vec{y} \in \mathcal{R}.$$

Then

$$\frac{1}{\prod\limits_{j=1}^{m} n_j} \sum_{\vec{i} \leq \vec{n}} \int_{[0,1]^l} f\left(\vec{s}\left(\vec{i}, \vec{n}\right), \vec{t}, \alpha\left(\vec{s}\left(\vec{i}, \vec{n}\right), \vec{t}\right), \xi(\vec{i}, \vec{n}, \vec{t})\right) d\vec{t} =$$

$$= \frac{1}{\prod\limits_{j=1}^{m} n_j} \sum_{\vec{i} \leq \vec{n}} \int_{[0,1]^l} \left(\left(\alpha_0\left(\vec{s}\left(\vec{i}, \vec{n}\right), \vec{t}\right) - \alpha\left(\vec{s}\left(\vec{i}, \vec{n}\right), \vec{t}\right)\right)^2 + \right.$$

$$+ 2\left(\alpha_0\left(\vec{s}\left(\vec{i}, \vec{n}\right), \vec{t}\right) - \alpha\left(\vec{s}\left(\vec{i}, \vec{n}\right), \vec{t}\right)\right) \xi(\vec{i}, \vec{n}, \vec{t})\right) d\vec{t} =$$

$$= \frac{1}{\prod_{j=1}^{m} n_j} \sum_{\vec{i} \leq \vec{n}} \int_{[0,1]^l} \left(\left(\alpha_0 \left(\vec{s} \left(\vec{i}, \vec{n} \right), \vec{t} \right) - \alpha \left(\vec{s} \left(\vec{i}, \vec{n} \right), \vec{t} \right) + \right.\right.$$

$$\left.\left. + \xi(\vec{i}, \vec{n}, \vec{t}) \right)^2 - \left(\xi(\vec{i}, \vec{n}, \vec{t}) \right)^2 \right) d\vec{t} = Q_{\vec{n}}(\alpha).$$

Denote

$$F(\alpha) = E \int_{[0,1]^l} \left(\left(\alpha_0(\vec{s}, \vec{t}) - \alpha(\vec{s}, \vec{t}) \right)^2 + 2 \left(\alpha_0(\vec{s}, \vec{t}) - \alpha(\vec{s}, \vec{t}) \right) \times \right.$$

$$\left. \times \xi(\vec{0}, \vec{1}, \vec{t}) \right) d\vec{s} \, d\vec{t} = \int_{[0,1]^l} \left(\alpha_0(\vec{s}, \vec{t}) - \alpha(\vec{s}, \vec{t}) \right)^2 d\vec{s} \, d\vec{t} \, .$$

It is evident that $F(\alpha) > F(\alpha_0) = 0$, $\alpha \neq \alpha_0$.

Now the theorem follows from Theorem 60. The proof is complete.

Let us now consider the asymptotic distribution of some functional of the least squares estimate of the unknown function $\alpha_{\vec{n}}$. For simplicity suppose that $m = 1$, $l = 1$. Denote $D = [0,1] \times [0,1]$. Then we need more restrictions on the set K and function α_0. Assume that the following condition is satisfied:

6) K is a set of all real functions $\alpha(s, t)$, which are defined on D and admit Furier expansion

$$\alpha(s, t) = \sum_{j,k=0}^{\infty} c_{jk}(\alpha) \, e^{2\pi i(js+kt)}, \tag{5.13}$$

where coefficients $c_{jk}(\alpha)$ satisfy such conditions

$$|c_{00}(\alpha)| \leq L, \qquad |c_{j0}(\alpha)| \, |j|^a \leq L,$$

$$|c_{0k}(\alpha)| \, |k|^b \leq L, \qquad |c_{jk}(\alpha)| \, |j|^a \, |k|^b \leq L, \tag{5.14}$$

where $a > 2$, $b > 2$, $L > 0$.

We will call $\alpha \in K$ an internal point of K if inequalities (5.14) hold for $\tilde{L} < L$.

Let us prove the following auxiliary statement.

Lemma 39 *Suppose that the continuous function $\alpha(s,t) : D \to \mathcal{R}$ satisfies the condition*

$$\left| \alpha(s_1,t) - \alpha(s_2,t) \right| \le c|s_1 - s_2| \tag{5.15}$$

for some $c > 0$ which does not depend on t, s_1, s_2. Then

$$\left| \frac{1}{n} \sum_{j=0}^{n} \int_0^1 \alpha(\frac{j}{n},t)\, dt - \iint_D \alpha(s,t)\, ds\, dt \right| \le \frac{c(n+1)}{n^2}. \tag{5.16}$$

In particulary

$$\left| \sum_{j=0}^{n} \int_0^1 \alpha(\frac{j}{n},t)\, dt \right| \le 2c$$

for the case when

$$\iint_D \alpha(s,t)\, ds\, dt = 0.$$

Proof. We have

$$\left| \frac{1}{n} \sum_{j=0}^{n} \int_0^1 \alpha(\frac{j}{n},t)\, dt - \iint_D \alpha(s,t)\, ds\, dt \right| = \left| \sum_{j=0}^{n} \int_{\frac{j}{n}}^{\frac{j+1}{n}} \int_0^1 \left[\alpha(\frac{j}{n},t)\, dt - \right. \right.$$

$$\left. - \alpha(s,t) \right]\, dt\, ds \left| \le c \sum_{j=0}^{n} \int_{\frac{j}{n}}^{\frac{j+1}{n}} \left| s - \frac{j}{n} \right|\, ds \le \frac{c(n+1)}{n^2}. \right.$$

The lemma is proved.

Corollary 3 *Lemma 39 remains true if instead of function $\alpha(s,t)$ we consider a peacely continuous function, which satisfies (5.15) on the intervals of continuity in the first argument. The lemma is true if consider the sum or the difference of such functions. In this case the constant c from (5.16) can be different from the constant in (5.15).*

Theorem 64 *Let* $E\Big(\xi(i,n,t)\Big)^4 \le \alpha$, $0 \le i \le n$, $n \in \mathcal{N}$, $t \in [0,1]$, *conditions* 1) $-$ 3), 6) *be satisfied,* $E\,\xi(i,n,t) = 0$ *and* α_0 *be an internal point of* K. *Then for any real function in* D *which can be introduced by*

$$\varphi(s,t) = \sum_{\substack{|j| \le \infty \\ |k| \le \infty}} |c_{jk}(\varphi)|\, e^{2\pi i(js+kt)}, \quad (s,t) \in D, \tag{5.17}$$

where $|c_{jk}(\varphi)|\,|j|^a\,|k|^b < \infty$, $a > 1$, $b > 1$ *and*

$$\iint_D \varphi^2(s,t)\,ds\,dt = 1,$$

for any $z \in \mathcal{R}$

$$\lim_{n \to \infty} P\left\{ \sqrt{n} \iint_D \varphi(s,t)\,[\alpha_n(s,t) - \alpha_0(s,t)]\,ds\,dt < z \right\} = \frac{1}{\sqrt{2\pi}\Delta} \int_{-\infty}^{z} e^{-\frac{u^2}{2\Delta^2}},$$

$$\Delta^2 = \int_0^1 \left[\iint_D R(s,t)\,\varphi(s,\lambda)\,\varphi(t,\lambda)\,ds\,dt \right] d\lambda,$$

$$R(s,t) = E\left(\xi(0,1,s)\,\xi(0,1,t) \right).$$

Proof. For all $\alpha \in K$ consider the representations

$$\alpha(s,t) = \theta(\alpha)\,\varphi(s,t) + \psi(s,t),$$

$$\psi(s,t) = \alpha(s,t) - \theta(\alpha)\,\varphi(s,t),$$

$$\iint_D \varphi(s,t)\,\psi(s,t)\,ds\,dt = 0, \qquad \iint_D \varphi(s,t)\,\alpha(s,t)\,ds\,dt = \theta(\alpha).$$

Let $\alpha_0(s,t) = \theta_0 \, \varphi(s,t) + \psi_0(s,t)$, $\alpha_n(s,t) = \theta_n \, \varphi(s,t) + \psi_n(s,t)$.

We shall note that $\psi_n(s,t)$, $(s,t) \in D$ is a random function. Theorem 63 implies that the sequence α_n converges uniformly on D with probability 1. Hence α_n is the minimum point of the functional $Q_n(\alpha)$, and satisfies the equation

$$\frac{1}{n} \sum_{j=0}^{n} \int_0^1 \left[x(j,n,u) - \alpha(\frac{j}{n},u) \right] \varphi(\frac{j}{n},u) \, du = 0$$

with probability, converging to 1, as $n \to \infty$.

That is why it is sufficient to consider the limit distribution of the value

$$\widetilde{\theta_n} = \frac{\displaystyle\sum_{j=0}^{n} \int_0^1 \left[x(j,n,u) - \psi_n(\frac{j}{n},u) \right] \varphi(\frac{j}{n},u) \, du}{\displaystyle\sum_{j=0}^{n} \int_0^1 \varphi^2(\frac{j}{n},u) \, du},$$

defined for sufficiently large n.

Let us show that the distribution of $\sqrt{n}\,(\widetilde{\theta_n} - \theta_0)$ converges weakly to the normal one. Let us introduce $\sqrt{n}\,(\widetilde{\theta_n} - \theta_0)$ as

$$\sqrt{n}(\widetilde{\theta_n} - \theta_0) = \eta_n + \zeta_n,$$

where

$$\eta_n = \frac{\dfrac{1}{\sqrt{n}} \displaystyle\sum_{j=0}^{n} \int_0^1 \left[\psi_0(\frac{j}{n},u) - \psi_n(\frac{j}{n},u) \right] \varphi(\frac{j}{n},u) \, du}{\dfrac{1}{n} \displaystyle\sum_{j=0}^{n} \int_0^1 \varphi^2(\frac{j}{n},u) \, du}, \tag{5.18}$$

$$\zeta_n = \frac{\dfrac{1}{\sqrt{n}} \displaystyle\sum_{j=0}^{n} \int_0^1 \varphi(\frac{j}{n},u) \, \xi\,(j,n,u) \, du}{\dfrac{1}{n} \displaystyle\sum_{j=0}^{n} \int_0^1 \varphi^2(\frac{j}{n},u) \, du}. \tag{5.19}$$

Because

$$\iint_D \varphi^2(u,v)\,du\,dv = 1, \qquad (5.20)$$

then the denominator in (5.18), (5.19) converges to 1, as $n \to \infty$.

Let us consider η_n. By the definitions of functions ψ_0 and ψ_n we have

$$\overline{\eta}_n = \frac{1}{\sqrt{n}} \sum_{j=0}^{n} \int_0^1 \left[\psi_0(\tfrac{j}{n}, u) - \psi_n(\tfrac{j}{n}, u) \right] \varphi(\tfrac{j}{n}, u)\,du =$$

$$= \frac{1}{\sqrt{n}} \sum_{j=0}^{n} \int_0^1 \left[\psi_0(\tfrac{j}{n}, u) - \psi_n(\tfrac{j}{n}, u) \right] \varphi(\tfrac{j}{n}, u)\,du -$$

$$- \sqrt{n} \iint_D [\psi_0(u,v) - \psi_n(u,v)] \varphi(u,v)\,du\,dv = \sqrt{n} \left\{ \frac{1}{n} \sum_{j=0}^{n} \int_0^1 \left[\alpha_0(\tfrac{j}{n}, u) - \right. \right.$$

$$\left. - \alpha_n(\tfrac{j}{n}, u) - (\theta_n - \theta_0)\varphi(\tfrac{j}{n}, u) \right] \varphi(\tfrac{j}{n}, u)\,du - \iint_D [\alpha_0(u,v) - \alpha_n(u,v) -$$

$$\left. - (\theta_n - \theta_0)\varphi(u,v)] \varphi(u,v)\,du\,dv \right\}. \qquad (5.21)$$

Because of (5.13), (5.14), (5.17) and (5.21) and Lemma 39

$$P\left\{ \lim_{n\to\infty} \overline{\eta}_n = 0 \right\} = 1. \qquad (5.22)$$

Hence from (5.18),(5.20), (5.22)

$$P\left\{ \lim_{n\to\infty} \eta_n = 0 \right\} = 1.$$

Consider now ζ_n. By virtue of (5.20) the distribution of ζ_n coincides asymptotically with the distribution of the value

$$\overline{\zeta}_n = \frac{1}{\sqrt{n}} \sum_{j=0}^{n} \int_0^1 \xi(j,n,t) \; \varphi(\frac{j}{n},t) \, dt.$$

The random variables $\int_0^1 \xi(j,n,t) \; \varphi(\frac{j}{n},t) \, dt$ are independent and

$$E\overline{\zeta}_n^2 = \frac{1}{n} \sum_{j=0}^{n} \iint_D R(s,t) \; \varphi(\frac{j}{n},t) \; \varphi(\frac{j}{n},s) \, ds \, dt,$$

$$\sum_{j=0}^{n} E \left[\frac{1}{\sqrt{n}} \int_0^1 \xi(j,n,t) \; \varphi(\frac{j}{n},t) \, dt \right]^4 \leq$$

$$\leq \frac{c_1}{n^2} \sum_{j=0}^{n} \int_{[0,1]^4} \left| E \prod_{i=1}^{4} \left(\xi(j,n,t) \; \varphi(\frac{j}{n},t_i) \right) \right| \prod_{i=1}^{4} dt_i \leq \frac{c_2}{n},$$

$$0 < c_1 < \infty, \quad 0 < c_2 < \infty.$$

Because of the central limit theorem the variable $\overline{\zeta}_n$ is asymptotically normal with mean 0 and variance

$$\int_0^1 \left[\iint_D R(s,t) \; \varphi(s,\lambda) \; \varphi(t,\lambda) \, ds \, dt \right] d\lambda.$$

The theorem is proved.

Let us note that we can obtain the asymptotical distribution of the integral functional of α_n with weaker restrictions on K.

Instead of condition 6) we will formulate the following conditions:

7) K is a set of all real functions defined in D and satisfying the conditions:

a)
$$\|\alpha(s,t)\| \le c_1, \quad 0 < c_1 < \infty,$$
where the constant c_1 does not depend on function $\alpha \in K$;

b)
$$|\alpha(\vec{z}_1) - \alpha(\vec{z}_2)| \le c_2 \| \vec{z}_1 - \vec{z}_2 \|$$
with a constant $c_2 > 0$, which does not depend on α, points $\vec{z}_1$ and $\vec{z}_2$, where $\vec{z} = (s,t)$;

8) for the function $\alpha_0 \in K \; \|\alpha_0\| < c_1$.

By Artsel theorem the set K is compact relatively to uniform convergence.

From conditions 7) and 8) we have that there exists such $\varepsilon > 0$ that for $|\theta| < \varepsilon$

$$\alpha_0 + \theta \in K.$$

Theorem 65 *Let the conditions of Theorem 64 be fulfilled, where instead of condition 6) conditions 7), 8) are satisfied. Then for all $z \in \mathcal{R}$*

$$\lim_{n \to \infty} P \left\{ \sqrt{n} \iint\limits_{D} [\alpha_n(u,v) - \alpha_0(u,v)] \, du \, dv < z \right\} = \frac{1}{\sqrt{2\pi}\Delta} \int\limits_{-\infty}^{z} e^{-\frac{u^2}{2\Delta^2}} \, du,$$

where

$$\Delta^2 = \iint\limits_{D} R(s,t) \, ds \, dt.$$

Proof. By virtue of Theorem 63 $\|\alpha_n - \alpha_0\| \to 0$, $n \to \infty$ with probability 1. Then with probability, converging to 1 as $n \to \infty$, $\alpha_n + \theta \in K$, $|\theta| < \varepsilon$ and α_n as the minimum point of $Q_n(\alpha)$ on K satisfies the equation

$$\sum_{j=0}^{n} \int\limits_{0}^{1} \left[x(j,n,u) - \alpha_n(\frac{j}{n}, u) \right] du = 0$$

with the same probability. Then the proof is similar to Theorem 64. The theorem is proved.

Now let us estimate the functional

$$\theta_0 = \iint\limits_{D} \varphi(u,v)\, \alpha_0(u,v)\, du\, dv$$

by the observations $\left\{x(j,n,u), \ 0 \le j \le n, \ u \in [0,1]\right\}$ without using an estimate of the unknown function α_0.

Consider the following estimate of the value θ_0:

$$\theta_n = \frac{1}{n} \sum_{j=0}^{n} \int_0^1 x(j,n,u)\, \varphi(\tfrac{j}{n},u)\, du.$$

We will formulate some assertions about the asymptotic behavior of the value θ_n.

Theorem 66 *Let conditions* 1) $-$ 3), 5) *be fulfilled,* $E\,\xi(j,n,u) = 0, \ 0 \le j \le n,$ $n \in \mathcal{N}, \ u \in [0,1],$ *the functions* $\alpha_0(u,v)$ *and* $\varphi(u,v)$ *be peacely continuous on* D *and satisfy on continuity intervals in the first argument to condition* (5.15). *Then*

$$P\left\{ \lim_{n\to\infty} \theta_n = \theta_0 \right\} = 1.$$

Proof. Consider $E\,\theta_n$:

$$E\,\theta_n = \frac{1}{n} \sum_{j=0}^{n} \int_0^1 \alpha_0(\tfrac{j}{n},u)\, \varphi(\tfrac{j}{n},u)\, du.$$

Because of corollary of Lemma 39

$$\left| \frac{1}{n} \sum_{j=0}^{n} \int_0^1 \alpha_0(\tfrac{j}{n},u)\, \varphi(\tfrac{j}{n},u)\, du - \iint\limits_{D} \alpha_0(u,v)\, \varphi(u,v)\, du\, dv \right| \le \frac{c}{n}.$$

Then $E\,\theta_n \to \theta_0, \ n \to \infty$. Let us consider the value

$$\frac{1}{n} \sum_{j=0}^{n} \int_0^1 \varphi(\tfrac{j}{n},u)\, \xi(j,n,u)\, du.$$

It is evident that values

$$\zeta_{jn} = \int\limits_0^1 \varphi(\frac{j}{n}, u)\, \xi(j, n, u)\, du$$

are independent, with zero means and restricted second moments. That is why by virtue of the strict law of large numbers

$$P\left\{ \lim_{n\to\infty} \frac{1}{n} \sum_{j=0}^n \zeta_{jn} = 0 \right\} = 1.$$

Analogously

$$P\left\{ \lim_{n\to\infty} \frac{1}{n} \sum_{j=0}^n \int\limits_0^1 \alpha_0(\frac{j}{n}, u)\, \varphi(\frac{j}{n}, u)\, du = 0 \right\} = 1.$$

The theorem is proved.

Theorem 67 *Let the conditions of Theorem 66 be fulfilled and*

$$E\Big(\xi(j, n, u) \Big)^4 \le L, \quad 0 \le j \le n, \quad n \in \mathcal{N}^m, \quad u \in [0, 1]^l, \quad L > 0.$$

Then for the value

$$\zeta_n = \sqrt{n}\,(\theta_n - \theta_0)$$

we have

$$\lim_{n\to\infty} P\left\{ \sqrt{n}\,[\theta_n - \theta_0] < z \right\} = \frac{1}{\sqrt{2\pi}\Delta} \int\limits_{-\infty}^z e^{-\frac{u^2}{2\Delta^2}}\, du, \quad z \in \mathcal{R},$$

where

$$\Delta^2 = \int\limits_0^1 \left[\iint\limits_D R(s, t)\, \varphi(s, \lambda)\, \varphi(t, \lambda)\, ds\, dt \right] d\lambda,$$

$$R(s, t) = E\left(\xi(0, 1, s)\, \xi(0, 1, t) \right).$$

Proof. Let us consider the value ζ_n:

$$\zeta_n = \sqrt{n}\left[\frac{1}{n}\sum_{j=0}^{n}\int_0^1 \alpha_0(\frac{j}{n},t)\,\varphi(\frac{j}{n},t)\,dt \;-\; \iint_D \varphi(u,v)\,\alpha_0(u,v)\,du\,dv\right] +$$

$$+ \frac{1}{\sqrt{n}}\sum_{j=0}^{n}\int_0^1 \varphi(\frac{j}{n},t)\,\xi\,(j,n,t)\,dt. \qquad (5.23)$$

Because of Lemma 39

$$\frac{1}{n}\sum_{j=0}^{n}\int_0^1 \alpha_0(\frac{j}{n},t)\,\varphi(\frac{j}{n},t)\,dt \;-\; \iint_D\;\iint_D \varphi(u,v)\,\alpha_0(u,v)\,du\,dv \to 0, \quad n\to\infty.$$

By virtue of the central limit theorem

$$\frac{1}{\sqrt{n}}\sum_{j=0}^{n}\int_0^1 \varphi(\frac{j}{n},t)\,\xi\,(j,n,t)\,dt \Longrightarrow \mathcal{N}(0,\Delta^2), \quad n\to\infty,$$

where sign " $\Longrightarrow$ " means weak convergence and Δ^2 is defined in formulation of the theorem.

Now (5.23) implies the proof of the theorem.

Let $\varphi_{st}(u,v) = \varphi_s(u)\,\varphi_t(v)$, where

$$\varphi_s(u) = \begin{cases} 1, & 0 \leq u \leq s \leq 1 \\ 0, & 0 < s < u \leq 1 \end{cases}.$$

We will define a random field

$$\theta_n(s,t) = \frac{1}{n}\sum_{j=0}^{n}\int_0^1 \varphi_{st}\left(\frac{j}{n},u\right)x(j,n,u)\,du$$

and let

$$\theta_0(s,t) = \int_0^s\int_0^t \alpha_0(u,v)\,du\,dv.$$

Theorem 68 *In conditions of Theorem 67 the finite-dimension distributions of the random field*

$$\zeta_n(s,t) = \sqrt{n}\,[\theta_n(s,t) - \theta_0(s,t)]$$

converges weakly to the finite-dimension distributions of Gaussian random field with mean 0 and correlation function

$$B(s_1,t_1,s_2,t_2) = \int\limits_0^{s_1}\int\limits_0^{s_2} R(u,v)\,du\,dv \cdot \min(t_1,t_2).$$

Proof. It is sufficiently to apply Theorem 67 to the random variable

$$\sqrt{n}\,[\theta_n - \theta_0] = \sum_{k=1}^m \lambda_k\,\zeta_n(s_k,t_k)$$

with the function

$$\varphi(u,v) = \sum_{k=1}^m \lambda_k\,\chi_{[0,s_k]}(u)\,\lambda_k\,\chi_{[0,t_k]}(v),$$

$$\chi_A(u) = \begin{cases} 1, & u \in A \\ 0, & u \bar\in A \end{cases}, \quad (u,v) \in D, \quad (s_i,t_i) \in D, \quad i = \overline{1,m}.$$

The theorem is proved.

All results about the least squares estimate are true for more complicated problem, when we have observations

$$x(j,n,t) = \alpha_0(\frac{j}{n},t)\,\eta(j,n,t) + \xi(j,n,t), \quad 0 \le t \le 1, \quad 0 \le j \le n, \quad n \in \mathcal{N}$$

and $\eta(j,n,t)$, $0 \le t \le 1$, $0 \le j \le n$, $n \in \mathcal{N}$. We need to estimate the function $\alpha_0 \in K$.

The least squares estimate is a minimum point for the function

$$\frac{1}{n}\sum_{j=0}^n \int_0^1 \left[x(j,n,t) - \alpha(\frac{j}{n},t)\,\eta(j,n,t) \right]^2 dt.$$

The family of random processes $\{\eta(j,n,t)\}$ satisfies the following conditions:

1) $\left\{\eta(j,n,t),\ 0 \le t \le 1\right\}, 0 \le j \le n$ for every n are independent real random processes with continuous traectories;

2) the finite-dimension distributions of random processes

$$\left\{\eta(j,n,t),\ 0 \le t \le 1\right\}, \quad 0 \le j \le n, \quad n \ge 1$$

do not depend on j, n;

3) random processes $\left\{(\eta(j,n,t))^2,\ 0 \le t \le 1\right\}, 0 \le j \le n$ are stationary in a wide sense and there exist $b, c_1 > 0$ such that

$$E\left(\eta(j,n,t)\right)^2 = b, \quad E\left(\eta(j,n,t)\right)^4 = c_1, \quad 0 \le t \le 1, \quad 0 \le j \le n,$$
$$n \in \mathcal{N};$$

4) the random processes $\left\{\xi(i,n,t),\ 0 \le t \le 1\right\}$, and $\left\{\eta(j,n,t),\ 0 \le t \le 1\right\}$ are independent for any i, j.

In that theorems, where we need $E\left(\xi(j,n,t)\right)^4 \le \alpha$ we demand that $E\left(\eta(j,n,t)\right)^8 \le \alpha$.

5.3 THE NONPARAMETRIC REGRESSION MODEL WITH OBSERVATIONS IN NODES OF A RECTANGLE

Let us now consider regression models where we estimate unknown function by finite number of observations.

Suppose that K is a compact subset of $C[0,1]^m$, $m \ge 1$; α_0 is a fixed but unknown function from K. We have observations

$$x(\vec{i}, \vec{n}) = \alpha\left(\vec{t}(\vec{i}, \vec{n})\right) + \xi(\vec{i}, \vec{n}), \quad \vec{i} \in \mathcal{Z}_+^m, \quad \vec{i} \le \vec{n},$$

where

$$\vec{n} = (n_j)_{j=1}^m \in \mathcal{N}^m, \quad \vec{i} = (i_j)_{j=1}^m, \quad \vec{t}(\vec{i}, \vec{n}) = \left(\frac{i_j}{n_j}\right)_{j=1}^m,$$

$$\left\{ \xi(\vec{i}, \vec{n}), \quad \vec{i} \leq \vec{n} \right\}$$

is a family of independent real random variables defined on a complete probabilistic space $(\Omega, \mathcal{G}, P)$ and satisfying the conditions:

1) $\xi(\vec{i}, \vec{n})$ has identical distributions for all $\vec{n} \in \mathcal{N}^m$, $\vec{i} \leq \vec{n}$;

2) zero is the single median of $\xi(\vec{i}, \vec{n})$ distribution;

3) $E\left(\xi(\vec{i}, \vec{n})\right)^2 \leq \infty$.

We need to estimate the function α_0. Consider the least modules estimate which is a minimum point of the function

$$F_{\vec{n}}(\alpha) = \frac{1}{\prod\limits_{j=1}^{m} n_j} \sum_{\vec{i} \leq \vec{n}} \left| x(\vec{i}, \vec{n}) - \alpha\left(\vec{t}\,(\vec{i}, \vec{n})\right) \right| =$$

$$= \frac{1}{\prod\limits_{j=1}^{m} n_j} \sum_{\vec{i} \leq \vec{n}} \left| \alpha_0\left(\vec{t}\,(\vec{i}, \vec{n})\right) - \alpha\left(\vec{t}\,(\vec{i}, \vec{n})\right) + \xi(\vec{i}, \vec{n}) \right|, \quad \alpha \in K. \quad (5.24)$$

Theorem 69 *For any $\vec{n} \in \mathcal{N}^m$, $\omega \in \Omega$, there exists at least one minimum point $\alpha_{\vec{n}} = \alpha_{\vec{n}}(\omega)$ for the functional (5.24), and for each $\vec{n}$ the function $\alpha_{\vec{n}}(\omega)$ $\omega \in \Omega$ can be chosen $\mathcal{G}_{\vec{n}}$-measurable, where*

$$\mathcal{G}_{\vec{n}} = \sigma\left\{ \xi\left(\vec{i}, \vec{n}\right), \quad \vec{i} \leq \vec{n} \right\}.$$

In this case

$$\|\alpha_{\vec{n}} - \alpha_0\| \to 0, \quad F_{\vec{n}}(\alpha_{\vec{n}}) \to E\left|\xi(\vec{0}, \vec{1})\right|, \quad \vec{n} \to \infty,$$

in probability.

If in condition 3) $E\left(\xi(\vec{i}, \vec{n})\right)^4 < \infty$ then

$$P\left\{ \|\alpha_{\vec{n}} - \alpha_0\| \to 0, \quad F_{\vec{n}}(\alpha_{\vec{n}}) \to E\left|\xi(\vec{0}, \vec{1})\right|, \quad \vec{n} \to \infty \right\} = 1.$$

Proof. It is easy to show that α_0 is a single minimum point of the functional

$$F(\alpha) = E \int_{[0,1]^m} \left| \alpha_0(\vec{t}) - \alpha(\vec{t}) + \xi(\vec{0}, \vec{1}) \right| d\vec{t}, \quad \alpha \in K.$$

This fact follows from Lemma 23 and Fubini theorem. Denote

$$f(\vec{t}, x, y) = \left| \alpha_0(\vec{t}) - x + y \right|, \quad \vec{t} \in [0,1]^m, \quad x \in \mathcal{R}, \quad y \in \mathcal{R}.$$

Then the proof of the theorem follows from Theorem 61.

Now we shall consider the least squares estimate, which is a minimum point of a functional

$$F_{\vec{n}}(\alpha) = \frac{1}{\prod\limits_{j=1}^{m} n_j} \sum_{\vec{i} \leq \vec{n}} \left(x(\vec{i}, \vec{n}) - \alpha\left(\vec{t}\,(\vec{i}, \vec{n}) \right) \right)^2 =$$

$$= \frac{1}{\prod\limits_{j=1}^{m} n_j} \sum_{\vec{i} \leq \vec{n}} \left(\alpha_0\left(\vec{t}\,(\vec{i}, \vec{n}) \right) - \alpha\left(\vec{t}\,(\vec{i}, \vec{n}) \right) + \xi(\vec{i}, \vec{n}) \right)^2, \quad \alpha \in K.$$

$$(5.25)$$

Theorem 70 *Assume that conditions* 1), 3) *are fulfilled, and* $E\xi(\vec{i}, \vec{n}) = 0$, $\vec{i} \leq \vec{n}$, $n \in \mathcal{N}^m$. *Then for any* $n \in \mathcal{N}^m$, $\omega \in \Omega$ *the functional* (5.24) *has at least one point of minimum* $\alpha_{\vec{n}} = \alpha_{\vec{n}}(\omega)$ *and for each* $\vec{n}$ *it can be chosen* $\mathcal{G}_{\vec{n}}$*-measurable. In this case*

$$\|\alpha_{\vec{n}} - \alpha_0\| \to 0, \quad Q_{\vec{n}}(\alpha_{\vec{n}}) \to 0, \quad \vec{n} \to \infty,$$

in probability, where

$$Q_{\vec{n}}(\alpha) = F_{\vec{n}}(\alpha) - \frac{1}{\prod\limits_{j=1}^{m} n_j} \sum_{\vec{i} \leq \vec{n}} \left(\xi(\vec{i}, \vec{n}) \right)^2.$$

If in condition 3) $E\left(\xi(\vec{i}, \vec{n}) \right)^4 < \infty$ *then*

$$P\left\{ \|\alpha_{\vec{n}} - \alpha_0\| \to 0, \quad Q_{\vec{n}}(\alpha_{\vec{n}}) \to 0, \quad \vec{n} \to \infty \right\} = 1.$$

This theorem follows from Theorem 61 in the same way as Theorem 63 follows from Theorem 60.

The results for least squares estimate are true when the observations are

$$x(\vec{i},\vec{n}) = \alpha_0 \left(\vec{t}\,(\vec{i},\vec{n})\right) \eta(\vec{i},\vec{n}) + \xi(\vec{i},\vec{n}), \quad \vec{i} \in Z_+^m, \quad \vec{i} \leq \vec{n},$$

and $\eta(\vec{i},\vec{n})$, $\vec{i} \leq \vec{n}$. We suppose that the following conditions are fulfilled:

1) $\left\{\eta(\vec{i},\vec{n}),\ \vec{i} \leq \vec{n}\right\}$ is a family of real independent and identically distributed random variables;

2) $E\left(\eta(\vec{i},\vec{n})\right)^4 < \infty.$

Let us now consider maximal likelihood estimate of unknown function α_0.

Suppose that for any $\vec{n} \in \mathcal{N}^m$, $\vec{i} \leq \vec{n}$ $\xi(\vec{i},\vec{n})$ has continuous distribution with probabilistic density g and $E\xi(\vec{i},\vec{n}) = 0$. The function g satisfies the conditions:

a) $g(u)$ is continuous and positive, $u \in \mathcal{R}$,

b) for the function

$$h(u) = \max_{|v| \leq 2c} |\ln g(u+v)|,$$

where $c = \max_{\alpha \in K} \|\alpha\|$, there exists $E h^2\left(\xi(\vec{0},\vec{1})\right) < \infty.$

The maximal likelihood estimate is a solution to the problem

$$\prod_{\vec{i} \leq \vec{n}} g\left(x(\vec{i},\vec{n}) - \alpha\left(\vec{t}\,(\vec{i},\vec{n})\right)\right) \to \max, \quad \alpha \in K. \qquad (5.26)$$

Problem (5.26) is equivalent to the problem

$$-\frac{1}{\prod_{j=1}^{m} n_j} \sum_{\vec{i} \leq \vec{n}} \ln g\left(x(\vec{i},\vec{n}) - \alpha\left(\vec{t}\,(\vec{i},\vec{n})\right)\right) = F_{\vec{n}}(\alpha) \to \min, \quad \alpha \in K.$$

$$(5.27)$$

Theorem 71 *For all $n \in \mathcal{N}^m$, $\omega \in \Omega$ there exists at least one solution $\alpha_{\vec{n}} = \alpha_{\vec{n}}(\omega)$ to the problem (5.26), and for any $\vec{n}$ it can be chosen $\mathcal{G}_{\vec{n}}$-measurable, where*

$$\mathcal{G}_{\vec{n}} = \sigma \left\{ \xi \left(\vec{i}, \vec{n} \right), \quad \vec{i} \leq \vec{n} \right\}.$$

In this case

$$\|\alpha_{\vec{n}} - \alpha_0\| \to 0, \quad F_{\vec{n}}(\alpha_{\vec{n}}) \to -E \ln g \left(\xi(\vec{0}, \vec{1}) \right), \quad \vec{n} \to \infty$$

in probability. If in condition b) for the function g we have $E\, h^4 \left(\xi(\vec{0}, \vec{1}) \right) < \infty$ then

$$P \left\{ \|\alpha_{\vec{n}} - \alpha_0\| \to 0, \quad F_{\vec{n}}(\alpha_{\vec{n}}) \to -E \ln g \left(\xi(\vec{0}, \vec{1}) \right), \quad \vec{n} \to \infty \right\} = 1.$$

Proof. Let us apply Theorem 61 to our model. Denote

$$f(\vec{t}, x, y) = -\ln g \left(\alpha_0(\vec{t}) - x + y \right), \quad \vec{t} \in [0,1]^m, x \in \mathcal{R}, y \in \mathcal{R}.$$

Then for problem (5.27)

$$F_{\vec{n}}(\alpha) = \frac{1}{\prod_{j=1}^{m} n_j} \sum_{\vec{i} \leq \vec{n}} f \left(\vec{t} \, (\vec{i}, \vec{n}), \alpha \left(\vec{t} \, (\vec{i}, \vec{n}) \right), \xi(\vec{i}, \vec{n}) \right).$$

Because of condition b) for the function g we have

$$E \left\{ \max_{\alpha \in K} \left(f \left(\vec{t}, \alpha(\vec{t}), \xi(\vec{0}, \vec{1}) \right) \right)^2 \right\} \leq \alpha, \quad \vec{t} \in [0,1]^m.$$

Denote

$$F(\alpha) = E \int_{[0,1]^m} f \left(\vec{t}, \alpha(\vec{t}), \xi(\vec{0}, \vec{1}) \right) d\vec{t} =$$

$$= E \int_{[0,1]^m} \left(-\ln g \left(\alpha_0(\vec{t}) - \alpha(\vec{t}) + \xi(\vec{0}, \vec{1}) \right) \right) d\vec{t}.$$

Suppose that $\alpha \neq \alpha_0$. Then for $\vec{t} \in S$, where S is some ball in $\mathcal{R}^m$, $S \subset [0,1]^m$, we have

$$E \ln g \left(\alpha_0(\vec{t}) - \alpha(\vec{t}) + \xi(\vec{0}, \vec{1}) \right) < E \ln g \left(\xi(\vec{0}, \vec{1}) \right).$$

This fact follows from Gensen inequality. Then $F(\alpha) > F(\alpha_0)$.

Now we have checked all conditions of Theorem 61. The theorem follows from Theorem 61.

5.4 THE PERIODICAL SIGNAL ESTIMATION BY OBSERVATION OF ITS MIXTURE WITH HOMOGENEOUS RANDOM FIELD

We will consider another nonparametric model where for proving of consistency of estimates Theorem 7 and Theorem 59 will be used.

Let a real random field $\{\xi(s,t), (s,t) \in \mathcal{R}^2\}$ be defined on the probabilistic space $(\Omega, \mathcal{F}, P)$. We need to estimate a function $\alpha_0(s,t)$ belonging to compact relatively to uniform convergence on square set K of 2π-periodical in every argument continuous functions by the observations

$$\Big\{ x(s,t) = \alpha_0(s,t) + \xi(s,t), \quad (s,t) \in D_T = [0,T]^2 \Big\}.$$

We will need the following conditions.

1. $\{\xi(s,t), (s,t) \in \mathcal{R}^2\}$ is a real homogeneous in a wide sense random field, its traectories are continuous with probability 1, $E\,\xi(s,t) = 0$, its correlation function is $r(s,t)$.

   We suppose also that a random field $\{|\xi(s,t)|, (s,t) \in \mathcal{R}^2\}$ is homogeneous in a wide sense with correlation function

   $$\widetilde{r}(s,t) = E\Big\{ \big(|\xi(s,t)| - E\,|\xi(0,0)|\big) \big(|\xi(0,0)| - E\,|\xi(0,0)|\big)\Big\}.$$

2. The functions r and $\widetilde{r}$ satisfy conditions

   $$\iint\limits_{D_T} |r(s,t)|\, ds\, dt = o\,(T^2), \quad T \to \infty,$$

$$\iint\limits_{D_T} |\tilde{r}(s,t)| \, ds \, dt = o \, (T^2), \quad T \to \infty.$$

3. The functions r and $\tilde{r}$ satisfy conditions

$$\iint\limits_{D_T} |r(s,t)| \, ds \, dt = O \, (T^{2-\delta}), \quad T \to \infty, \quad \delta > 0,$$

$$\iint\limits_{D_T} |\tilde{r}(s,t)| \, ds \, dt = O \, (T^{2-\delta}), \quad T \to \infty, \quad \delta > 0. \qquad (5.28)$$

4. Let K be a set of real periodical in both arguments with period 2π functions with the representation

$$\alpha(s,t) = \sum_{k,l=-\infty}^{\infty} c_{kl}(\alpha) \, e^{i(ks+lt)},$$

where Furier coefficients

$$c_{kl}(\alpha) = \frac{1}{4\pi^2} \int\limits_{0}^{2\pi} \int\limits_{0}^{2\pi} \alpha(s,t) \, e^{i(ks+lt)} \, ds \, dt$$

satisfy conditions

$$|c_{00}(\alpha)| \leq L, \qquad |c_{k0}(\alpha)| \, |k|^a \leq L,$$
$$\qquad\qquad\qquad\qquad\qquad\qquad L > 0, \quad a > 2, \quad b > 2.$$
$$|c_{0l}(\alpha)| \, |l|^b \leq L, \qquad |c_{kl}(\alpha)| \, |k|^a \, |l|^b \leq L,$$

Let us consider the least squares estimate $\alpha_T(s,t)$ of unknown function α_0. It is defined as an element $\alpha_T \in K$ for which we have the relation

$$\iint\limits_{D_T} [\alpha_T(s,t) - x(s,t)]^2 \, ds \, dt = \min_{\alpha \in K} \iint\limits_{D_T} [\alpha(s,t) - x(s,t)]^2 \, ds \, dt.$$

Because of the compactness of K relatively to uniform convergence on the plane and continuity of traectories $\{\xi(s,t),\ (s,t) \in \mathcal{R}^2\}$ the function $\alpha_T(s,t)$ exists. Let us show that it is unique. Suppose that $\alpha_T^1(s,t)$ and $\alpha_T^2(s,t)$ are two optimal estimates being got by the least squares method. Then for the convex set K we have

$$2 \iint\limits_{D_T} \left[\alpha_T^1(s,t) - x(s,t)\right]^2 ds\,dt + 2 \iint\limits_{D_T} \left[\alpha_T^2(s,t) - x(s,t)\right]^2 ds\,dt =$$

$$= \iint\limits_{D_T} \left[\alpha_T^1(s,t) - \alpha_T^2(s,t)\right]^2 ds\,dt + 4 \iint\limits_{D_T} \left[\frac{\alpha_T^1(s,t) + \alpha_T^2(s,t)}{2} - \right.$$

$$\left. - x(s,t)\right]^2 ds\,dt \geq \iint\limits_{D_T} \left[\alpha_T^1(s,t) - \alpha_T^2(s,t)\right] ds\,dt +$$

$$+4 \iint\limits_{D_T} \left[\alpha_T^1(s,t) - x(s,t)\right]^2 ds\,dt.$$

Hence

$$\iint\limits_{D_T} \left[\alpha_T^1(s,t) - \alpha_T^2(s,t)\right]^2 ds\,dt = 0,$$

i.e. $\alpha_T^1(s,t) = \alpha_T^2(s,t)$. Then $\alpha_T^1 = \alpha_T^2$ if K is convex and $T \geq 2\pi$.

As for another estimates considered before it is evident that $\{\alpha_T(s,t),\ (s,t) \in \mathcal{R}^2\}$ is a separable measurable field.

Denote

$$\|\alpha\|^2 = \frac{1}{4\pi^2} \int\limits_0^{2\pi}\int\limits_0^{2\pi} \alpha^2(s,t)\,ds\,dt.$$

The following assertion takes place.

Theorem 72 *Let conditions 1, 2 be fulfilled and K be compact relatively to uniform convergence on the plane set of 2π-periodical in every argument continuous functions. Then*

$$\sup_{(s,t)\in\mathcal{R}^2} \left| \alpha_T(s,t) - \alpha_0(s,t) \right| \to 0, \quad T \to \infty$$

in probability.

Proof. Denote

$$Q_T(\alpha) = \frac{1}{T^2} \iint_{D_T} \left[\alpha(s,t) - x(s,t) \right]^2 ds\, dt - \frac{1}{T^2} \iint_{D_T} \xi(s,t)\, ds\, dt.$$

It is evident that $Q_T(\alpha)$, $\alpha \in K$ has its minimal value when $\alpha = \alpha_T$. Besides,

$$\lim_{T\to\infty} \frac{1}{T^2} \iint_{D_T} \left[\alpha(s,t) - \alpha_0(s,t) \right]^2 ds\, dt = \|\alpha - \alpha_0\|^2.$$

Let us show that

$$a_T(\alpha) = \frac{1}{T^2} \iint_{D_T} \left[\alpha(s,t) - \alpha_0(s,t) \right] \xi(s,t)\, ds\, dt$$

converges to 0 as $T \to \infty$ in mean squares sense. We have

$$E\left| a_T(\alpha) \right|^2 = \frac{1}{T^4} \iint_{D_T} \iint_{D_T} \left[\alpha(s_1,t_1) - \alpha_0(s_1,t_1) \right] \left[\alpha(s_2,t_2) - \alpha_0(s_2,t_2) \right] \times$$

$$\times\, r\left(s_1 - s_2,\, t_1 - t_2 \right) ds_1\, dt_1\, ds_2\, dt_2 \leq 4\, c_0 \frac{1}{T^2} \int_{-T}^{T} \int_{-T}^{T} \left(1 - \frac{|u|}{T} \right) \left(1 - \frac{|v|}{T} \right) \times$$

$$\times \left| r(u,v) \right| du\, dv \to 0, \quad T \to \infty,$$

$$c_0 = \max_{\alpha \in K} \max_{(s,t) \in \mathcal{R}^2} |\alpha(s,t)|.$$

That is why as $T \to \infty$

$$Q_T(\alpha) = \frac{1}{T^2} \iint\limits_{D_T} \Big[\alpha(s,t) - \alpha_0(s,t) \Big]^2 ds\,dt - 2\,a_T(\alpha) \to \|\alpha - \alpha_0\|^2 \quad (5.29)$$

in probability. Then condition 3) of Theorem 59 is fulfilled with the function $\Phi(\alpha; \alpha_0) = \|\alpha - \alpha_0\|^2$. To check condition 4) of Theorem 59 consider for $\gamma > 0$

$$\zeta_T(\alpha, \gamma) = \sup_{(\widetilde{\alpha}\,:\,\|\alpha - \widetilde{\alpha}\| < \gamma)} \Big| Q_T(\widetilde{\alpha}) - Q_T(\alpha) \Big| =$$

$$= \sup_{(\widetilde{\alpha}\,:\,\|\alpha - \widetilde{\alpha}\| < \gamma)} \frac{1}{T^2} \iint\limits_{D_T} \Big[\widetilde{\alpha}(s,t) - \alpha(s,t) \Big] \times$$

$$\times \Big[\widetilde{\alpha}(s,t) + \alpha(s,t) - 2\,\alpha_0(s,t) - 2\,\xi(s,t) \Big]\, ds\,dt.$$

It is easy to see that

$$\zeta_T(\alpha, \gamma) \leq 2\gamma \left[2c_0 + \frac{1}{T^2} \iint\limits_{D_T} \Big| \xi(s,t) \Big|\, ds\,dt \right],$$

and by condition 2

$$\frac{1}{T^2} \iint\limits_{D_T} \Big| \xi(s,t) \Big|\, ds\,dt \to E \Big| \xi(0,0) \Big|, \quad T \to \infty$$

in mean squares sense. Then if $c(\gamma) = 4\gamma \left(2c_0 + E\Big| \xi(0,0) \Big| \right)$ we have

$$\lim_{T \to \infty} P\Big\{ \zeta_T(\alpha, \gamma) < c(\gamma) \Big\} = 1. \quad (5.30)$$

Hence

$$\sup_{(s,t) \in \mathcal{R}^2} \Big| \alpha_T(s,t) - \alpha_0(s,t) \Big| \to 0, \quad T \to \infty$$

in probability. The proof is complete.

Theorem 73 *Let conditions 1, 2 hold only for the field $\xi(s,t)$ (not for $\left|\xi(s,t)\right|$) and condition 4 be fulfilled. Then*

$$\sup_{(s,t)\in\mathcal{R}^2}\left|\alpha_T(s,t)-\alpha_0(s,t)\right|\to 0,\quad T\to\infty$$

in probability.

Proof. Relation (5.29) is proved in the same way as in Theorem 72. For proving relation (5.30) we have

$$\zeta_T(\alpha,\gamma)\le 4c_0\gamma+2\sup_{(\tilde{\alpha}:\|\alpha-\tilde{\alpha}\|<\gamma)}\sum_{k,l=-\infty}^{\infty}\left|c_{kl}(\tilde{\alpha})-c_{kl}(\alpha)\right|\times$$

$$\times\left|\frac{1}{T^2}\iint_D e^{i(ks+lt)}\xi(s,t)\,ds\,dt\right|\le 4c_0\gamma+4\sum_{k,l=-\infty}^{\infty}{}'|k|^{-a}|l|^{-b}\times$$

$$\times\left|\frac{1}{T^2}\iint_{D_T}e^{i(ks+lt)}\xi(s,t)\,ds\,dt\right|,$$

where $\sum'$ denotes that for $k=0$ or $l=0$ instead of $|k|^{-a},|l|^{-b}$ we have 1. Because of condition 2

$$E\left\{\sum_{k,l=-\infty}^{\infty}{}'|k|^{-a}|l|^{-b}\left|\frac{1}{T^2}\iint_{D_T}e^{i(ks+lt)}\xi(s,t)\,ds\,dt\right|\right\}\le\sum_{k,l=-\infty}^{\infty}{}'|k|^{-a}|l|^{-b}\times$$

$$\times\left[\frac{1}{T^2}\int_{-T}^{T}\int_{-T}^{T}\left(1-\frac{|s|}{T}\right)\left(1-\frac{|t|}{T}\right)|r(s,t)|\,ds\,dt\right]^{1/2}\to 0,\quad T\to\infty.$$

Then if $c(\gamma)=5c_0\gamma$ then

$$\lim_{T\to\infty}P\left\{\zeta_T(\alpha,\gamma)<c(\gamma)\right\}=1.$$

The proof is complete.

Theorem 74 *Assume that conditions* $1, 3, 4$ *are fulfilled. Then*

$$P\left\{ \lim_{T\to\infty} \sup_{(s,t)\in\mathcal{R}^2} |\alpha_T(s,t) - \alpha_0(s,t)| = 0 \right\} = 1. \qquad (5.31)$$

Proof. Let us show that

$$P\left\{ \lim_{T\to\infty} a_T(\alpha) = 0 \right\} = 1.$$

From (5.28) we have

$$E\,|a_T(\alpha)|^2 \le \frac{c}{T^\delta}.$$

Let p be a fixed integer number for which $\delta_p > 1$. Then by virtue of Borel-Cantelli Lemma for $T(n) = n^p$

$$P\left\{ \lim_{T\to\infty} a_{T(n)} = 0 \right\} = 1. \qquad (5.32)$$

Let $T \in [T(n), T(n+1)]$. Then $a_T \le a_{T(n)} + \zeta_n$, where

$$\zeta_n \;=\; \zeta_{n1} + \zeta_{n2},$$

$$\zeta_{n1} \;=\; \left| \frac{1}{(T(n))^2} \int\limits_{T(n)}^{T(n+1)} \int\limits_{0}^{T(n+1)} [\alpha(s,t) - \alpha_0(s,t)]\,\xi(s,t)\,ds\,dt \right|,$$

$$\zeta_{n2} \;=\; \left| \frac{1}{(T(n))^2} \int\limits_{0}^{T(n)} \int\limits_{T(n)}^{T(n+1)} [\alpha(s,t) - \alpha_0(s,t) - 2]\,\xi(s,t)\,ds\,dt \right|.$$

Similarly to (5.32)

$$P\left\{ \lim_{n\to\infty} \zeta_{n1} = 0 \right\} = 1, \qquad (5.33)$$

$$P\left\{ \lim_{n\to\infty} \zeta_{n2} = 0 \right\} = 1. \qquad (5.34)$$

From (5.32) - (5.34) we have

$$P\left\{ \lim_{T\to\infty} a_T(\alpha) = 0 \right\} = 1.$$

Similarly

$$P\left\{ \lim_{T\to\infty} \frac{1}{T^2} \iint\limits_{D_T} |\xi(s,t)|\, ds\, dt = E\,|\xi(0,0)| \right\} = 1.$$

Denote

$$c(\gamma) = 4\gamma\,(2c_0 + E\,|\xi(0,0)|).$$

Then

$$P\left\{ \overline{\lim_{n\to\infty}}\, \zeta_T(\alpha,\gamma) < c(\gamma) \right\} = 1. \tag{5.35}$$

All conditions of Theorem 7 are fulfilled. The proof is complete.

Theorem 75 *Let condition 4 be fulfilled and conditions 1, 3 be fulfilled only for the field $\xi(s,t)$. Then condition (5.31) takes place.*

Proof. The process of proving is similar to that one in the previous theorem. To check condition (5.35) we need the relation

$$P\left\{ \lim_{T\to\infty} \sideset{}{'}\sum_{k,l=-\infty}^{\infty} |k|^{-a}\,|l|^{-b}\, \frac{1}{T^2} \left| \iint\limits_{D_T} e^{i(ks+lt)}\,\xi(s,t)\, ds\, dt \right| = 0 \right\} = 1.$$

It is easy to show that the last relation follows from Borel-Cantelli lemma. The proof is complete.

REFERENCES

[1] ALEXANDRYAN R.A. AND MIRZAHANYAN A.A. (1979) *General Topology*, Vischaya Shkola: Moscow (in Russian)

[2] ANDERSON T.W. (1958) *An Introduction to Multivariate Statistical Analysis*, Wiley

[3] ANDERSON T.W. (1971) *The Statistical Analysis of Time Series*, Wiley

[4] BEREZANSKII YU.M., US G.F. AND SHEFTEL Z.G. (1990) *Functional Analysis*, Vishcha Shkola: Kiev (in Russian)

[5] BILLINGSLEY P. (1968) *Convergence of Probability Measures*, Wiley

[6] BOROVKOV A.A. (1986) *Probability Theory*, Nauka: Moscow

[7] CRAMER H. (1946) *Mathematical Methods of Statistics*, University of Stockholm

[8] CRAMER H. AND LEADBETTER M.R. (1967) *Stationary and Related Stochastic Process: Sample Function Properties and Their Applications*, Wiley

[9] DOOB J.L. (1953) *Stochastic Processes*, Wiley

[10] DOROGOVTSEV A.YA. (1974) *Asymptotic Properties of Estimators of the Parameter of a Signal that is Amplitude Modulated by a Random Process*, Dop. Akad. Nauk Ukrain. R.S.R.A, 4, 680-684 (in Ukrainian)

[11] DOROGOVTSEV A.YA. (1975) *On Limit Behavior of One Estimate of the Regression Function*, Teor. Veroyatnost. Matem. Statist., 13, 38-46 (in Russian)

[12] DOROGOVTSEV A.YA. (1975) *Properties of Parameters Estimates in One Linear Regression Model*, Visn. Kyiv. Univers.: Ser. Matem. Mekhan. 17, 3-12 (in Ukrainian)

240

[13] DOROGOVTSEV A.YA. (1976) *Notes On Properties of One Non-parametric Maximal Likelihood Estimate*, Matem. Sbornik, Naukova Dumka: Kiev, 262-268 (in Russian)

[14] DOROGOVTSEV A.YA. (1976) *On One Statement Useful for Estimates Consistency Proving*, Teor. Veroyatnost. Matem. Statist., 14, 34-41 (in Russian)

[15] DOROGOVTSEV A.YA. (1982) *The Theory of Estimation of Random Processes Parameters*, Vishcha Shkola: Kiev (in Russian)

[16] DOROGOVTSEV A.YA. (1992) *Consistency of Least Squares Estimator of an Infinite Dimensional Parameter*, Siber Math. J., 33, 65-69 (in Russian)

[17] DOROGOVTSEV A.YA. (1993) *On Asymptotic Normality of Least Squares Estimator of an Infinite Dimensional Parameter*, Ukrain. Math. J., 45, 44-53 (in Russian)

[18] DOROGOVTSEV A.YA. AND IVANOV A.V. (1975) *Properties of the Parameters Estimate in Nonlinear Regression Models*, Izdat. Obshch. Znaniye: Kiev (in Russian)

[19] DOROGOVTSEV A.YA. AND KNOPOV P.S. (1976) *On Consistency of Estimates of the Continuous in the Domain Function by Observations of Its Values in a Finite Set of Points with Random Errors*, Dokl. Akad. Nauk Ukrain. S.S.R.A, 12, 1065-1069 (in Russian)

[20] DOROGOVTSEV A.YA. AND KNOPOV P.S. (1976) *On the Properties of an Estimator of the Parameters of a Regression on a Nonstationary Gaussian Random Field*, Teor. Veroyatnost. Matem. Statist., 15, 54-68 (in Russian)

[21] DOROGOVTSEV A.YA. AND KNOPOV P.S. (1977) *Asymptotic Properties of One Nonparametric Function of Two Variables*, Teor. Sluch. Processov, 5, 27-35 (in Russian)

[22] DOTSENKO S.V. (1974) *On Estimation of the Values of Physical Fields of an Ocean by Results of Their Measurements on Rectilinear Traectories*, Morsk. Gydrophys. Issled., 2, 148-161 (in Russian)

[23] DUPAČOVA J. (1987) *Stochastic Programming with Incomplete Information: A Survey of Results on Post Optimization and Sensitivity Analysis*, Optimization, 18, 507-532

[24] DUPAČOVA J. AND WETS R.J-B. (1988) *Asymptotic Behavior of Statistical Estimators and Optimal Solutions for Stochastic Optimization Problems*, Ann.Statist., 16, 1517-1549

[25] ERMOLIEV YU.M. (1970) *On Optimal Control of Random Processes*, Kibernetika, 2, 18-29 (in Russian)

[26] ERMOLIEV YU.M. (1976) *Methods of Stochastic Programming*, Nauka: Moscow (in Russian)

[27] ERMOLIEV YU.M., GULENKO V.P. AND TSARENKO T.I. (1978) *The Finite Difference Method in Optimal Control Problems*, Naukova Dumka: Kiev (in Russian)

[28] FELLER W. (1966) *An Introduction to Probability Theory and Its Applications*, VOL.II, Wiley

[29] FORTUS M.I. AND YAGLOM A.M. (1963) *An Estimate for the Coefficients of a Linear Combination of Given Functions in the Presence of Noise with a Rational Spectrum*, Probl. Peredachi Informatsii, 14, 136-150 (in Russian)

[30] VAN DE GEER S. (2000) *Applications of Empirical Process Theory*, Cambridge Series in Statistical and Probabilistic Mathematics, 6, Cambridge University Press

[31] GIKHMAN I.I. AND SKOROKHOD A.V. (1971) *Stochastic Processes Theory*, VOL.I, Nauka: Moscow (in Russian)

[32] GIKHMAN I.I. AND SKOROKHOD A.V. (1977) *Introduction to Stochastic Processes Theory*, Nauka: Moscow (in Russian)

[33] GILCHRIST R. AND PORTIDES G. (1994) *Using GLIM4 to Estimate the Tunding Constant for Huber's M-Estimate of Location*, Springer-Verlag: Hiedelberg

[34] GNEDENKO B.V. (1969) *Course in Probability Theory*, Nauka: Moscow (in Russian)

[35] GRECHKA G.P. AND DOROGOVTSEV A.YA. (1976) *Asymptotic Properties of the Periodogram Estimate of Frequency and Amplitude of a Harmonic Oscillation*, Vychisl. Prikl. Matem., 28, 18-31 (in Russian)

242

[36] GRENANDER U. (1950) *Stochastic Processes and Statistical Inference*, Almqvist & Wiksells Boktryckeri AB: Stockholm

[37] GRENANDER U. (1954) *On the Estimation of Regression Coefficients in the Case of an Autocorrelated Disturbance*, Ann. Math. Statist., 25, N2 252-272

[38] GRENANDER U. AND ROSENBLATT M. (1956) *Statistical Analysis of Stationary Time Series*, Uppsala

[39] HAMPEL F.R., RONCHETTI E.M., ROUSSEEUW P.J. AND STAHEL W.J. (1986) *Robust Statistics: Approach Based on Influence Functions*, Wiley

[40] HANNAN E.J. (1971) *Non-Linear Time Series Regression*, J. Appl. Prob., 8, 767-780

[41] HANNAN E.J. (1973) *The Estimation of Frequency*, J. Appl. Prob., 10, 510-519

[42] HANNAN E.J. (1976) *Multiple Time Series*, Wiley

[43] HEBLE M.P. (1961) *A Regression Problem Concerning Stationary Processes*, Trans. Amer. Math. Soc., 99, N2, 350-371

[44] HOLEVO A.S. (1971) *On Asymptotic Normality of Estimates of Regression Coefficients*, Teor. Veroyatnost. Primen., 16, N4, 724-728 (in Russian)

[45] HUBER P.J. (1967) *The Behavior of Maximum Likelihood Estimates under Nonstandard Conditions*, Proc. 5th Berkeley Symp. on Mathematical Statistics and Probability. I, University of California Press: Berkeley, 221-234

[46] HUBER P.J. (1981) *Robust Statistics*, Wiley

[47] HUSU A.P., VITTENBERG YU.R. AND PAL'MOV V.A. (1975) *Roughness of the Surface*, Nauka: Moscow (in Russian)

[48] IBRAGIMOV I.A. AND HAS'MINSKII R.Z. (1973) *Approximation of Statistical Estimates by Summes of Independent Random Variables*, Dokl. Akad. Nauk S.S.S.R., 210, N6, 1273-1276 (in Russian)

[49] IBRAGIMOV I.A. AND HAS'MINSKII R.Z. (1974) *An Estimate of a Signal Parameter in Gaussian White Noise*, Probl. Peredachi Informatsii, 10, N1, 39-59 (in Russian)

[50] IBRAGIMOV I.A. AND HAS'MINSKII R.Z. (1977) *On Estimation of an Infinite Dimensional Parameter in Gaussian White Noise,* Dokl. Akad. Nauk S.S.S.R., 236, N5, 1053-1055 (in Russian)

[51] IBRAGIMOV I.A. AND HAS'MINSKII R.Z. (1977) *One Problem of Statistical Estimation in Gaussian White Noise,* Dokl. Akad. Nauk S.S.S.R., 236, N6, 1300-1302 (in Russian)

[52] IBRAGIMOV I.A. AND HAS'MINSKII R.Z. (1979) *Asymptotic Theory of Estimation,* Nauka: Moscow (in Russian)

[53] IBRAGIMOV I.A. AND HAS'MINSKII R.Z. (1980) *On Nonparametric Estimation of Regression,* Dokl. Akad. Nauk S.S.S.R., 252, N4, 780-784 (in Russian)

[54] IBRAGIMOV I.A. AND LINNIK YU.V. (1965) *Independent and Stationary Sequences of Random Variables,* Nauka: Moscow (in Russian)

[55] IBRAGIMOV I.A. AND ROZANOV YU.A. (1970) *Gaussian Stochastic Processes,* Nauka: Moscow (in Russian)

[56] IBRAMHALILOV I.S. AND SKOROKHOD A.V. (1980) *Consistency Estimators of Random Processes Parameters,* Naukova Dumka: Kiev (in Russian)

[57] IVANOV A.V. (1979) *A Solution of the Problem of Detection of Hidden Periodicities,* Teor. Veroyatnost. Matem. Statist., 20, 44-60 (in Russian)

[58] IVANOV A.V. (1984) *On Consistency and Asymptotic Normality of Least Moduli Estimator,* Ukrain. Math. J, 36, 267-272 (in Russian)

[59] IVANOV A.V. (1991) *Estimation Theory of Nonlinear Regression Models Parameters,* Doctor of Sciences Thesis: Institute of Mathematics of the Ukrainian Academy of Sciences: Kiev (in Russian)

[60] IVANOV A.V. (1997) *Asymptotic Theory of Nonlinear Regression,* Kluwer Academic Publishers: Dordrecht

[61] IVANOV A.V. AND LEONENKO N.N. (1989) *Statistical Analysis of Random Fields,* Kluwer Academic Publishers: Dordrecht

[62] JENNRICH R.I. (1969) *Asymptotic Properties of Non-Linear Least Squares Estimators*, Ann. Math. Statist., 40, 633-643

[63] JURBENKO I.G. (1987) *Stationary and Homogeneous Stochastic Systems Analysis*, Moscow University: Moscow (in Russian)

[64] KARMANOV V.G. (1975) *Mathematical Programming*, Nauka: Moscow (in Russian)

[65] KASITSKAYA E.J. (1989) *Properties of Least Squares Estimates for Regression of Gaussian Homogeneous Isotropic Fields*, Mathematical Methods of Analysis and Optimization of Complicated Systems Operating in the Conditions of Indeterminacy, Collection of Scientific Works, Institute of Cybernetics of the Ukrainian S.S.R. Academy of Sciences: Kiev, 22-27 (in Russian)

[66] KASITSKAYA E.J. (1990) *Approximation of the Solution of the Stochastic Programming Problem with the Disturbance which is a Homogeneous Random Field*, Mathematical Methods for Taking of Decisions in the Conditions of Indeterminacy, Collection of Scientific Works, Institute of Cybernetics of the Ukraininan S.S.R. Academy of Sciences: Kiev, 23-27 (in Russian)

[67] KASITSKAYA E.J. AND KNOPOV P.S. (1990) *Asymptotic Behavior of Empirical Estimates in Stochastic Programming Problems*, Dokl. Akad. Nauk S.S.S.R., 315, N2, 279-281 (in Russian)

[68] KASITSKAYA E.J. AND KNOPOV P.S. (1991) *On Convergence of Empirical Estimates in Stochastic Optimization Problems*, Kibernetika, 2, 104-107, 112 (in Russian)

[69] KASITSKAYA E.J. AND KNOPOV P.S. (1991) *About One Approach to the Nonlinear Estimating Problem*, Proc. 6th USSR-Japan Symposium on Probability Theory and Mathematical Statistics, Kiev, World Scientific: Singapore, 151-157

[70] KING A. (1986) *Asymptotic Behavior of Solutions in Stochastic Optimization: Nonsmooth Analysis and the Derivation of Non-Normal Limit Distributions*, Dissertation: University of Washington

[71] KING A. (1988) *Asymptotic Distribution for Solutions in Stochastic Optimization and Generalized M-Estimation,* (Preprint), International Institute for Applied Systems Analysis: Laxenburg, Austria

[72] KNOPOV P.S. (1976) *On Some Estimates of Nonlinear Parameters for the Stochastic Field Regression,* Teor. Veroyatnost. Matem. Statist., 14, 67-74 (in Russian)

[73] KNOPOV P.S. (1976) *On Asymptotic Properties of Some Nonlinear Regression Estimates,* Teor. Neroyatnost. Matem. Statist., 15, 73-82 (in Russian)

[74] KNOPOV P.S. (1979) *On Asymptotic Behavior of Periodogram Estimates of Parameters in Nonlinear Regression Models,* Dokl. Akad. Nauk Ukrain. S.S.R.A, 11, 942-945 (in Russian)

[75] KNOPOV P.S. (1980) *On Some Problems of Nonparametric Estimation of Stochastic Fields,* Dokl. Akad. Nauk Ukrain. S.S.R.A, 9, 79-82 (in Russian)

[76] KNOPOV P.S. (1981) *Optimal Estimators of Parameters of Stochastic Systems,* Naukova Dumka: Kiev (in Russian)

[77] KNOPOV P.S. (1984) *Estimation of the Unknown Parameters of an almost Periodic Function in the Presence of Noise. I,* Kibernetika, 6, 83-87 (in Russian);

(1985) *Optimal Estimation of the Unknown Parameters of an almost Periodic Function in the Presence of Noise. II,* Kibernetika, 3, 82-85 (in Russian)

[78] KNOPOV P.S. (1986) *Methods of Nonlinear Identification and Recognition for Stochastic Systems with Distributed Parameters,* Institute of Cybernetics of the Ukrainian S.S.R. Academy of Sciences: Kiev (in Russian)

[79] KNOPOV P.S. (1988) *On One Approach to Solution of Stochastic Optimization Problems,* Kibernetika, 4, 126-127 (in Russian)

[80] KNOPOV P.S. (1997) *Asymptotic Properties of One Class of M-Estimates,* Kibernet. Sist. Anal., 4, 10-27 (in Russian)

[81] KNOPOV P.S. (1997) *On a Nonstationary Model of M - Estimators with Discrete Time,* Teor. Yimovirnost. Matem. Statist., 57, 60-66 (in Ukrainian)

[82] KNOPOV P.S. (1998) *Estimates of Parameters for Nonidentic-ally Distributed Random Variables*, Teor. Yimovirnost. Matem. Statist., 58, 38-44 (in Ukrainian)

[83] KNOPOV P.S. AND KASITSKAYA E.J.(1989) *On Asymptotic Behavior of Nonlinear Parameters of Random Functions*, Proc. 4th International Conference on Probability Theory and Mathematical Statistics, Vilnus, 65-70 (in Russian)

[84] KNOPOV P.S. AND KASITSKAYA E.J. (1989) *Asymptotic Properties of Least Squares Estimates of Gaussian Regression for Random Fields*, Kibernetika, 5, 64-68 (in Russian)

[85] KNOPOV P.S. AND KASITSKAYA E.J. (1993) *On Some Problems of Identification of Nonlinear Regression Parameters in a Discrete Case*, Kibernet. Vychislit. Tekhn, 97, 11-15 (in Russian)

[86] KNOPOV P.S. AND KASITSKAYA E.J. (1994) *Least Modules Method in Identification Models with Discrete Time*, Kibernet. Vychislit. Tekhn, 101, 80-86 (in Russian)

[87] KNOPOV P.S. AND KASITSKAYA E.J. (1995) *Properties of Empirical Estimates in Stochastic Optimization and Identification Problems*, Ann. Oper. Res., 56, 225-239

[88] KNOPOV P.S. AND KASITSKAYA E.J. (1999) *Consistency of Least Squares Estimates for Parameters of the Gaussian Regression Model*, Kibernet. Sist. Anal., 1, 21-26 (in Russian)

[89] KOLMOGOROV A.N. AND FOMIN S.V. (1976) *Elements of the Theory of Functions and Functional Analysis*, Nauka: Moscow (in Russian)

[90] KORKHIN A.S. (1985) *On Some Properties of Estimates of Regression Parameters under Apriori Restrictions in the Form of Inequalities*, Kibernetika, 6, 106-114 (in Russian)

[91] KOROLJUK V.S., PORTENKO N.I., SKOROKHOD A.V. AND TURBIN A.F. (1985) *Handbook of Probability Theory and Mathematical Statistics*, Nauka: Moscow (in Russian)

[92] KUKUSH A.G. (1989) *Asymptotic Properties of the Estimator of a Nonlinear Regression Infinite Dimensional Parameter*, Math. Today, 5, 84-105 (in Russian)

[93] KUKUSH A.G. (1995) *Asymptotic Properties of the Estimator of Infinite Dimensional Parameters of Random Processes*, Doctor of Sciences Thesis: Kiev Mathematical Institute (in Russian)

[94] KUTOYANTS YU.A. (1980) *Random Processes Parameters Estimation*, Akademiya Nauk Armyanskoi S.S.R.: Yerevan (in Russian)

[95] LEONENKO N.N. (1999) *Limit Theorems for Random Fields with Singular Spectrum*, Kluwer Academic Publishers: Dordrecht

[96] LEONENKO N.N. AND YADRENKO M.I. (1975) *Central Limit Theorem for Homogeneous and Isotropic Random Fields*, Dokl. Akad. Nauk Ukrain. S.S.R.A, 4, 314-316 (in Russian)

[97] LEONENKO N.N. AND YADRENKO M.I. (1978) *On Estimates of Regression Coefficients for a Homogeneous Random Field*, Ukrain. Mat. Zh., 6, 749-756 (in Russian)

[98] LIPTSER R.SH. AND SHIRYAYEV A.N. (1974) *Statistics of Random Processes*, Nauka: Moscow (in Russian)

[99] LJUNG L. (1987) *System Identification: Theory for the User*, Prentic-Hall

[100] LOEVE M. (1963) *Probability Theory*, Van Nostrand: Princeton, NJ

[101] LYASHKO I.I., BOYARCHUK A.K., GUY YA.G. AND KALAYDA A.F. (1983) *Mathematical Analysis*. I, Vishcha Shkola: Kiev (in Russian)

[102] LYASHKO I.I., BOYARCHUK A.K., GUY YA.G. AND KALAYDA A.F. (1985) *Mathematical Analysis*. II, Vishcha Shkola: Kiev (in Russian)

[103] MANN H.B. AND MORANDA P.B. (1954) *On the Efficiency of the Least Squares Estimates of Parameters in the Ornshtein-Uhlenbeck Process*, Sankhya, 13, 351-358

[104] NEMIROVSKII A.S., POLJAK B.T. AND TSYBAKOV A.V. (1984) *Estimates of Signals by Nonparametric Maximal Likelihood Method*, Probl. Peredachi Informatsii, 20, N3, 29-46 (in Russian)

[105] *Numerical Techniques for Stochastic Optimization*, ED. ERMOLIEV YU.M. AND WETS R. J-B. (1988), Springer: Berlin

[106] PFANZAGL J. (1969) *On the Measurability and Consistency of Minimum Contrast Estimates*, Metrika, 14, 249-272

[107] PFANZAGL J. AND WEFELMEYER W. (1985) *Asymptotic Expansions for General Statistical Models*, Lecture Notes in Statistics, 31, Springer-Verlag: Berlin

[108] PFLUG G. (1996) *Optimization of Stochastic Models*, Kluwer Academic Publishers: Dordrecht

[109] PISARENKO V.F. AND ROZANOV YU.A. (1963) *On Some Problems for Stationary Processes Leading to Integral Equations Related with Winner-Hoppf Equation*, Probl. Peredachi Informatsii, 14, 113-135 (in Russian)

[110] POLJAK B.T.(1983) *Introduction to Optimization*, Nauka: Moscow (in Russian)

[111] PRAKASA RAO B.L.S. (1987) *Asymptotic Theory of Statistical Inference*, Wiley

[112] PROKHOROV YU.V. (1956) *Convergence of Stochastic Processes and Limit Theorems of Probability Theory*, Teor. Veroyatnost. Primen., 1, N2, 177-238 (in Russian)

[113] PSHENICHNY B.N. AND MARCHENKO D.N. (1967) *On One Approach to Finding of the Global Minimum*, Teor. Optimal. Reshen., 2, 3-12 (in Russian)

[114] RAO C.R. (1965) *Linear Statistical Inference and Its Applications*, Wiley

[115] ROZANOV YU.A. (1964) *The Editor Supplement in the book: Hennan A., Analysis of Time Series*, Nauka: Moscow (in Russian)

[116] ROZANOV YU.A. (1967) *Stationary Random Processes*, Holden-Day

[117] SALINETTI G. AND WETS R.J-B. (1986) *On the Convergence in Distribution of Measurable Multifunctions (Random Sets), Normal Integrands, Stochastic Processes and Stochastic Infima*, Math. Oper. Res., 11, 385-419

[118] SCHMETTERER L. (1974) *Introduction to Mathematical Statistics*, Springer-Verlag: Berlin

[119] SEREBROVSKII M.G. AND PERVOZVANSKII A.A. (1965) *Detection of Hidden Periodicities*, Nauka: Moscow (in Russian)

[120] SHAPIRO A. (1989) *Asymptotic Properties of Statistical Estimators in Stochastic Programming*, Ann. Statist., 17, 841-858

[121] SHAPIRO A. (1991) *Asymptotic Analysis of Stochastic Programs*, Ann. Oper. Res., 30, 169-186

[122] STRIEBEL CH.T. (1961) *Efficient Estimation of a Regression Parameter for Certain Second Order Processes*, Ann. Math. Statist., 32, N4, 1299-1313

[123] TSE-PEI CHIANG (1959) *On the Estimation of Regression Coefficients of a Continuous Parameter Time Series*, Teor. Veroyatnost. Primen., 4, N4, 405-423

[124] VAPNIK V.N. (1979) *Estimation of Dependence Based on Empirical Data*, Nauka: Moscow (in Russian)

[125] VASILIEV F.P. (1981) *Methods for Solving of Extremal Problems*, Nauka: Moscow (in Russian)

[126] WALD A. (1949) *Note on the Consistency of the Maximum Likelihood Estimate*, Ann. Math. Statist., 20, N2, 595-601

[127] WALKER A.M. (1973) *On the Estimation of a Harmonic Component in a Time Series with Stationary Dependent Residuals*, Adv. Appl. Prob., 5, 217-241

[128] WETS R.J-B. (1983) *Stochastic Programming: Solution Techniques and Approximation Schemes*, Mathematical Programming. The State of the Art, Springer-Verlag: Berlin, 566-603

[129] YADRENKO M.I. (1971) *Statistical Problems for Isotropic Random Fields*, Proc. 8th Summer Mathematical School, Institute of Mathematics of the Ukrainian Academy of Sciences: Kiev, 237-282 (in Russian)

[130] YADRENKO M.I. (1980) *The Spectral Theory of Random Fields*, Vishcha Shkola: Kiev (in Russian)

[131] YAGLOM A.M. (1952) *An Introduction to the Theory of Stationary Random Functions*, Uspekhi Matemat. Nauk, 7, N5 (51), 1-168 (in Russian)

[132] YAGLOM A.M. (1987) *Correlation Theory of Stationary and Related Random Functions*, I, II, Series in Statistics, Springer-Verlag: Berlin

[133] YU-CZUY CHAN (1965) *Estimate of Regression Coefficients for Stochastic Processes with a Continuous Parameter and a Stationary Disturbance*, Acta Sci. Natur. Fudan., 10, N2-3, 101-112

[134] YUDIN D.B. (1979) *Problems and Methods of Stochastic Programming*, Sov'etskoye Radio: Moscow (in Russian)